Crystal Chemistry of Zinc, Cadmium and Mercury

Crystal Chemistry of Zinc, Cadmium and Mercury

Special Issue Editor

Matthias Weil

MDPI • Basel • Beijing • Wuhan • Barcelona • Belgrade

MDPI

Special Issue Editor
Matthias Weil
TU Wien
Institute for Chemical Technologies and Analytics
Getreidemarkt 6/164-SC
1060 Vienna, Austria

Editorial Office
MDPI
St. Alban-Anlage 66
4052 Basel, Switzerland

This is a reprint of articles from the Special Issue published online in the open access journal *Crystals* (ISSN 2073-4352) from 2017 to 2018 (available at: https://www.mdpi.com/journal/crystals/special_issues/cadmium_mercury)

For citation purposes, cite each article independently as indicated on the article page online and as indicated below:

LastName, A.A.; LastName, B.B.; LastName, C.C. Article Title. *Journal Name* **Year**, *Article Number, Page Range.*

ISBN 978-3-03897-652-3 (Pbk)
ISBN 978-3-03897-653-0 (PDF)

Contents

About the Special Issue Editor

Matthias Weil, 1990–1996: Chemistry studies at Justus-Liebig-University of Gießen (Germany); 1996–1999: Dissertation "Preparation and crystal chemistry of phosphates and arsenates of Hg and Cd, and of ultraphosphates MP_4O_{11} with M = Fe, Zn and Cd" at Justus-Liebig-University of Gießen under the supervision of Prof. Dr. Reginald Gruehn, 1999–2004: Postdoc TU Wien (Austria); 2004: Assistant Professor TU Wien; 2006: Habilitation and Venia Docendi, TU Wien; 2006: Associate Professor TU Wien.

Preface to "Crystal Chemistry of Zinc, Cadmium and Mercury"

The closed-shell $nd^{10}(n+1)s^2$ (n = 3,4,5) electronic configuration of group 12 elements (zinc, cadmium, mercury) defines their respective physical and chemical behaviours. Hence, for some properties, zinc and cadmium compounds resemble their alkaline earth congeners (likewise with closed-shell configurations) rather than their transition metal relatives, whereas mercury and its compounds are unique among all metals, which, to a certain extent, can be attributed to relativistic effects. All these features are also reflected in the characteristic crystal chemistry of the zinc triad elements and their compounds, including simple inorganic salts, alloys, compounds with inorganic framework structures, compounds with molecular structures or coordination polymers and hybrid compounds.

Matthias Weil

Special Issue Editor

crystals

MDPI

Article

Theoretical Analysis of Elastic, Mechanical and Phonon Properties of Wurtzite Zinc Sulfide under Pressure

Melek Güler * and Emre Güler

Department of Physics, Hitit University, 19030 Corum, Turkey; eguler71@gmail.com
* Correspondence: mlkgnr@gmail.com; Tel.: +90-3642277000; Fax: +90-3642277005

Academic Editor: Matthias Weil
Received: 18 April 2017; Accepted: 31 May 2017; Published: 4 June 2017

Abstract: We report for the first time the application of a mixed-type interatomic potential to determine the high-pressure elastic, mechanical, and phonon properties of wurtzite zinc sulfide (WZ-ZnS) with geometry optimization calculations under pressures up to 12 GPa. Pressure dependency of typical elastic constants, bulk, shear, and Young moduli, both longitudinal and shear wave elastic wave velocities, stability, as well as phonon dispersions and corresponding phonon density of states of WZ-ZnS were surveyed. Our results for the ground state elastic and mechanical quantities of WZ-ZnS are about experiments and better than those of some published theoretical data. Obtained phonon-related results are also satisfactory when compared with experiments and other theoretical findings.

Keywords: ZnS; wurtzite; elastic; mechanical; phonon

1. Introduction

The computational modeling of materials has been a successful and rapid tool to address the unclear points of physical interest. In addition, predicting good elastic and thermodynamic properties of materials is a necessary demand for present-day solid state science and industry. In particular, these properties at high pressure and temperature are important for the development of modern technologies [1–5].

As emphasized in the authors' recent work, wide-band-gap II–VI semiconductors such as ZnX (X: S, O, Se, Te) and CdX (X: S, Se, Te) are remarkable materials for the design of high-performance opto-electronic devices including light-emitting diodes and laser diodes in the blue or ultraviolet region [6].

According to its crystallographic structure, ZnS compounds can crystallize in either zinc blende (ZB) crystal structure with space group F-43m or wurtzite (WZ) crystal structure with the P6$_3$mc space group at ambient conditions [7–10].

To date, there have been a number of studies performed especially for the ground state (T = 0 K and P = 0 GPa) structural, elastic, thermodynamic, and optical properties of ZB and WZ crystal structures of ZnS compounds—not only with experiments, but also with theoretical works because of their technological importance. In 2004, Wright and Gale [11] reported new interatomic potentials to model the structures and stabilities of the ZB and WZ ZnS phases in which their theoretical results are within experiments. Later, in 2008, Bilge et al. [8], performed ab initio calculations based on projector augmented wave pseudo potential (PAW). They employed generalized gradient approximation (GGA) of density functional theory (DFT) to investigate the ground state mechanical and elastic properties of ZB, rock salt (NaCl), and WZ phases of ZnS. They concluded that the mechanical properties of ZnS under high pressure are quite different from those at ambient conditions. At the same time, Rong et al. [10] reported the pressure dependence of the elastic properties of ZB and WZ crystals of ZnS by the GGA within the plane-wave pseudopotential (PWP) of DFT and found reasonable results which are consistent with former experimental and theoretical data. Further, in 2009, Cheng et al. [12] conducted an experimental and theoretical study on first and second-order Raman scattering of ZB and WZ ZnS and found satisfactory experimental and theoretical results. Afterwards, in 2012, Grünwald et al. [9] established transferable pair potentials for ZnS crystals to accurately describe the ground state properties of ZB and WZ phases of ZnS compounds. Recently, Yu et al. [13] performed DFT calculations by using both the local density approximation (LDA) and GGA for the exchange-correlation potential and calculated the phonon dispersion curves and the phonon density of states of WZ ZnS in which the calculated values display good agreement with earlier experimental and theoretical data. In another recent study, Ferahtia et al. [14] published the first-principles plane-wave-based pseudopotential method calculations of the structural, elastic, and piezoelectric parameters of WZ ZnS and concluded a reasonable degree of agreement between their results and data available in the literature.

The above recent and increasing theoretical efforts on ZnS motivated us to perform this work by addressing the lesser-known high-pressure elastic, mechanical, and phonon properties of WZ ZnS with a different method. In contrast to the above applied methods and interatomic potentials of literature, this is the first report regarding the application of a mixed-type potential to determine the mentioned high-pressure quantities of WZ ZnS with geometry optimization calculations. During calculations, we concentrated on the pressure behavior of five independent elastic constants, bulk, shear, and Young moduli, elastic wave velocities, mechanical stability conditions, and phonon properties of WZ ZnS under pressures between 0 GPa and 12 GPa at T = 0 K.

The following part of the article provides details of our theoretical calculations with the employed interatomic potential in Section 2. In Section 3, we compare our results with the available experimental results and other theoretical data. Finally, Section 4 presents the main findings of this work in the conclusions.

2. Details of Theoretical Calculations

The most significant feature of materials modeling is the choice of the proper interatomic potentials that sufficiently and accurately describe the physical properties of the concerned problem. Simple empirical potentials are such modelling tools for materials, since they can yield reasonable results. These potentials can also provide good explanations of the defect energies, surface energies, elastic properties, and mechanical aspects of oxides [15], fluorides [16] and many other materials [17,18]. Most of these potentials include Coulomb interactions, short-range pair interactions, and ionic polarization treated by the shell model of Dick and Overhauser [19]. The sum of the Coulomb terms, short-range interactions, and ionic polarization expresses the total energy for these potentials. If we presume that the electron cloud of an ion is simulated by a massless shell charge q_s and the nucleus by a core of charge q_c, then the total charge becomes $q = q_s + q_c$. In the shell model interatomic potential approach, a harmonic force with spring constant K couples the core and the shell of the ion. So, for modelling the short-range pair interactions acting between the shells, we can use a typical Buckingham potential, as presented in Equation (1):

$$V_{ij}^{Buckingham} = A \exp\left(-\frac{r_{ij}}{\rho}\right) - C r_{ij}^{-6} \tag{1}$$

The first part of Equation (1) indicates the Born–Mayer term, whereas the second part represents the Van der Waals energies. Further, in our work, we applied the original form of a mixed-type interatomic potential of Hamad et al. [20] for short-range interactions, which incorporates the Buckingham and Lennard–Jones 9–6 potentials form as in Equation (2):

$$V_{ij}^{short} = A \exp\left(-\frac{r_{ij}}{\rho}\right) + B r_{ij}^{-9} - C r_{ij}^{-6} \tag{2}$$

We also considered the semi-covalent nature of ZnS by using a three-body potential for S-Zn-S angel as in its original form [20] and expressed by Equation (3):

$$V_{ijk} = \left(\frac{1}{2}\right) K_{TB} \left(\theta_{ijk} - \theta_0\right)^2 \tag{3}$$

In Equation (3), θ_0 and K_{TB} show the equilibrium constant angle between S-Zn-S and fitting constant of Hamad et al. [20], respectively. Lastly, sulphur anion polarizibility treated by the shell model of Reference [19] can be written as in Equation (4):

$$V_{ij}^{core-shell} = \left(\frac{1}{2}\right) K r_{ij}^2 \tag{4}$$

where r_{ij} describes the core–shell separation, and K indicates the spring constant. Although extra details of presently employed potential and its parameterization procedure can be found in Reference [20], Table 1 lists the original potential parameters of Reference [20] used in our calculations.

Table 1. Mixed type theoretical interatomic potential used in this work. Parameters of the applied potential were taken from its original reference [20].

Mixed Potential Parameters from Ref. [20]				
General potential	A (eV)	ρ (Å)	B (eV. Å^9)	C (eV. Å^6)
Zn-S	213.20	0.475	664.35	10.54
S-S	11,413	0.153	0.0	129.18
Three-body potential	θ_0 (degree)		K_{TB} (ev.rad^{-2})	
Zn core-S Shell-S Shell	109. 47		0.778	
Spring potential			K (eV. Å^{-2})	
S core-S Shell	27.690			
Ion charges	Charge (e)			
Zn core	2.000			
S core	1.357			
S Shell	−3.357			

All theoretical calculations in this work were carried out with the General Utility Lattice Program (GULP) 4.2 molecular dynamics code [21,22]. This versatile code allows the concerned structures to be optimized at constant pressure (all internal and cell variables are included) or at constant volume (unit cell remains frozen). To avoid the constraints, constant pressure optimization was applied to the geometry of WZ ZnS cell with the Newton–Raphson method based on the Hessian matrix calculated from the second derivatives. The cell geometry of WZ ZnS was assigned as $a = b = 3.811$ Å, $c = 6.234$ Å, $u = 0.375$, and $\alpha = \beta = 90°$ and $\gamma = 120°$ with space group $P6_3mc$. During the present geometry optimization calculations, the Hessian matrix was recursively updated with the BFGS [23–26] algorithm. After setting the necessities for the geometry optimization of WZ ZnS, we devised multiple runs at zero Kelvin temperature and checked the pressure ranges between 0 GPa and 12 GPa in steps of 3 GPa. Further, phonon and associated properties of WZ ZnS were also addressed after constant pressure geometry optimization calculations as a function of pressure within the quasiharmonic approximation under zero Kelvin temperature, as implemented in GULP code. It is possible to capture the phonon density of states (PDOS) and dispersions for a material after specifying a shrinking factor with GULP phonon computations. Further, phonons are described by calculating their values at points in reciprocal space within the first Brillouin zone of the given crystal. To achieve the Brillouin zone integration and obtain the PDOS, we have used a standard and reliable scheme developed by Monkhorts and Pack [27] with $8 \times 8 \times 8$ *k-point* mesh.

3. Results and Discussion

Figure 1 indicates the density behavior of WZ ZnS under pressure. As is well known, the density of many materials exhibits clear increments under pressure [6,15,17,18,28]. This is also the case for WZ ZnS under pressure, as seen in Figure 1. The minimum and maximum values of the density of WZ ZnS was found to be 4.08 g/cm^3 and 4.59 g/cm^3 for 0 GPa and 12 GPa, respectively, at zero temperature.

The elastic constants of materials not only provide precise and essential information about the materials, but also explain many mechanical and physical properties. Once the elastic constants are determined, one may get a deeper insight into the stability of the concerned material [6,15,17,18,28]. These constants can be also helpful to predict the properties of materials (i.e., interatomic bonding, equation of state, and phonon spectra). They also link to several thermodynamic parameters, such as the specific heat, thermal expansion, Debye temperature, Grüniesen parameter, etc. However, in general, elastic constants derived from the total energy calculations correspond to single-crystal elastic properties. So, the Voigt–Reuss–Hill approximation is a confident scheme for polycrystalline materials [6,15,17,18,28]. To obtain the accurate values of elastic constants and other analyzed parameters of WZ ZnS, the Voigt–Reuss–Hill values were taken into account for this research. For wurtzite crystal structures, five independent elastic constants exist as: $C_{11}, C_{12}, C_{13}, C_{33},$ and C_{44} [28].

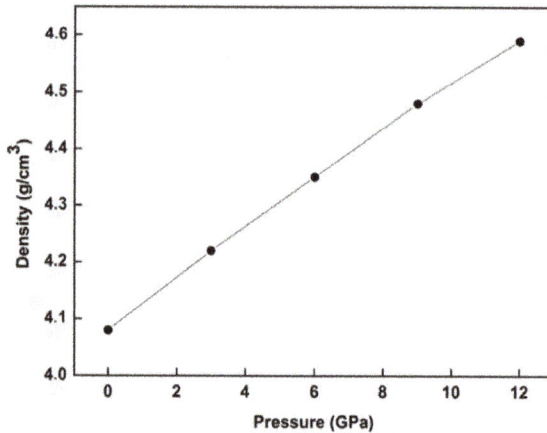

Figure 1. Density behavior of wurtzite (WZ) ZnS under pressure.

The plot in Figure 2 shows our results for $C_{11}, C_{12}, C_{13}, C_{33}$ and C_{44} elastic constants of WZ ZnS under pressures between 0 GPa and 12 GPa. All obtained elastic constants of WZ ZnS increase with the applied pressure except C_{33} and C_{44}. Beyond these increments, a closer examination of Figure 2 reveals that the magnitudes of the elastic constants are in the range of: $C_{33} > C_{11} > C_{12} > C_{13} > C_{44}$. Both the range of elastic constants and slight decrement of C_{44} constant under pressure mimic the DFT findings of Tan et al. [7]. However, our results for the ground state parameters of WZ ZnS are obviously better than those of Tan et al. [7] as well as Wright and Gale [11], and are much closer to the experimental results of Neumann et al. [29]. Table 2 gives a numerical comparison of present results for the elastic constants and other calculated quantities of WZ ZnS with some previously published experimental and theoretical data.

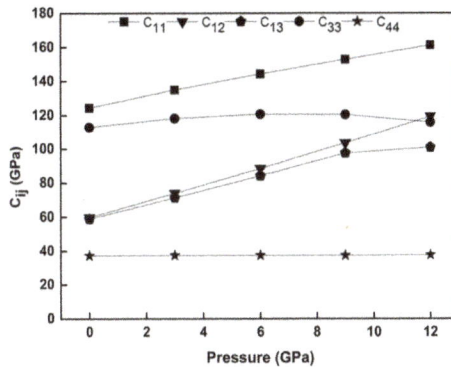

Figure 2. Elastic constants of WZ ZnS under pressure.

Table 2. Comparing our results with previous experimental and theoretical data for the considered parameters of WZ ZnS under zero pressure and temperature. * Bulk modulus experimental value was taken from Reference [30].

Parameter	Exp [29]	Present	Other Theoretical	
			Ref [8]	Ref [14]
C_{11} (GPa)	124.2	124.4	115.6	135.4
C_{12} (GPa)	60.1	59.8	48.9	65.8
C_{13} (GPa)	45.5	58.9	37.1	51.6
C_{33} (GPa)	140.0	113.0	132.5	160.7
C_{44} (GPa)	28.6	37.3	27.2	32.4
B (GPa)	75.8 *	79.5	68.5	84.9
E (GPa)		75.3		
G (GPa)		33.4		
V_S (km/s)		2.86		
V_L (km/s)		5.51		

From a stability outlook, the proverbial Born mechanical stability condition for a hexagonal structure also holds for wurtzite crystal, and must satisfy [28]:

$$C_{44} > 0, \; C_{11} > C_{12} \text{ and } (C_{11} + 2C_{12})C_{33} - 2C_{13}^2 > 0$$

The present results for the obtained elastic constants of WZ ZnS satisfy the mechanical stability condition (Figure 2 and Table 2), and this result points out that WZ ZnS is mechanically stable at 0 K temperature and 0 GPa pressure.

The bulk modulus (B) is the only elastic constant of a material that conveys much information about the bonding strength. Moreover, it is a measure of the material's resistance to external deformation, and occurs in many formulae describing diverse mechanical–physical characteristics. The shear modulus (G), however, portrays the resistance to shape change caused by a shearing force. In addition to B and G, Young's modulus (E) is the resistance to uniaxial tensions. These three distinct moduli (B, G, and E) are other valuable parameters for identifying the mechanical properties of materials [6,15,17,18,28].

Figure 3 represents the pressure behavior of B, G, and E of WZ ZnS for the entire pressure range. From the prevalent physical definition of bulk modulus ($B = \Delta P / \Delta V$) is expected an increment for B because of its direct proportion to applied pressure. Hence, bulk modulus of WZ ZnS exemplifies a straight increment in Figure 3. Contrary to sharp increment in B, E and G moduli have slight decrements under pressure again similar to the DFT findings of Tan et al. [7]. Apart from the pressure behavior of these three moduli, Table 2 also summarizes another numerical comparison for B with former experimental and theoretical data. The present result for the bulk modulus of WZ ZnS with 79.5 GPa approximately agrees with experimental values and is better than other DFT results, as seen in Table 2.

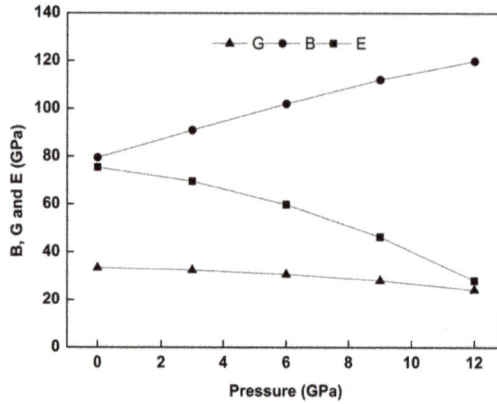

Figure 3. Elastic moduli (B, G, and E) behavior of WZ ZnS versus pressure.

The adjectives "brittle" and "ductile" signify the two distinct mechanical characters of solids when they are exposed to stress. Ductile and brittle features of materials play a key role during the manufacturing of materials [6,15,17,18,28]. So, we also evaluated the ductile (brittle) nature of WZ ZnS under pressure. In general, brittle materials are not deformable or are less deformable before fracture, whereas ductile materials are very deformable before fracture. At this point, the Pugh ratio is a determinative limit for the ductile (brittle) behavior of materials, and has popular use in literature. If the B/G ratio is about 1.75 and higher, the material is accepted to be ductile; otherwise, the material becomes brittle [6,15,17,18,28]. After a careful evaluation, we determined B/G values changing from 2.38 to 4.95, with a monotonous increment between P = 0 GPa and P = 12 GPa at zero temperature, respectively, for WZ ZnS (as seen in Figure 4). So, this result manifests the ductile character of WZ ZnS in both ground state and under pressure. As another result, all values of the B/G are higher than 1.75 and increase with pressure, which suggests that pressure can improve ductility (Figure 4).

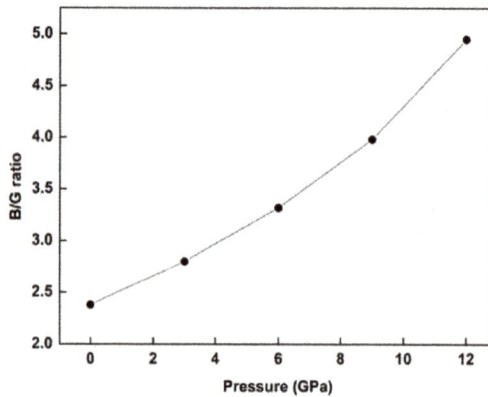

Figure 4. B/G ratio of WZ ZnS against pressure.

In solids, low temperature (T = 0 K in our case) acoustic modes can trigger the vibrational excitations. Depending on this fact, two typical (longitudinal and shear) elastic waves exist. The velocity V_L represents the longitudinal elastic wave velocity and V_S denotes shear wave velocity [6,15,17,18,28]. Figure 5 shows the behavior of V_L and V_S elastic wave velocities for WZ ZnS as a function of pressure. V_L has a substantial increment compared to V_S, which is the most observed circumstance for many materials.

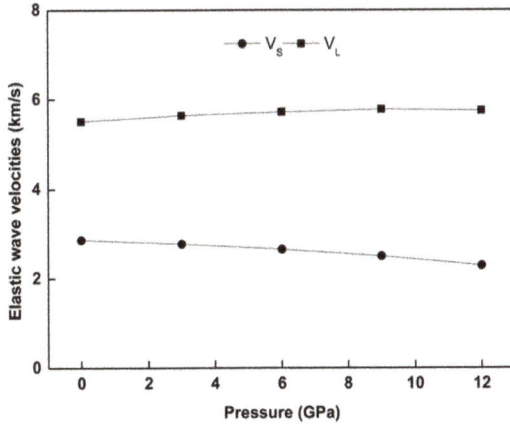

Figure 5. Behavior of longitudinal (V_L) and shear wave (V_S) velocities of WZ ZnS under pressure.

Figure 6 shows the phonon dispersion of WZ ZnS along the chosen Γ–A path. In addition, Table 3 compares the numerical values of the zone–center (Γ points) phonon frequencies of this work with earlier experimental and theoretical data of WZ ZnS. As seen in Table 3, the agreement with experiment is quite good, especially for A1 (TO), E1 (TO) and A2 (TO), E2 (TO) phonon modes and competing with recent GGA and LDA DFT data of Reference [14].

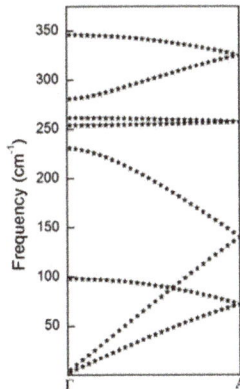

Figure 6. Phonon dispersion curve of WZ ZnS at zero temperature along the chosen Γ–A path.

Table 3. Comparing our results with previous experimental and theoretical data for the phonon frequencies (cm^{-1}) of WZ ZnS under zero pressure and temperature.

Symmetry	Exp [31] (cm^{-1})	Present (cm^{-1})	Ref [13]	
			GGA (cm^{-1})	LDA (cm^{-1})
E_2 (low)	72	98	68	69
B_1 (low)		230	186	199
A1 (TO)	273	279	257	290
E1 (TO)	273	253	257	293
E_2 (high)	286	260	261	297
B_1 (high)		279	318	347
A1 (LO)	351	345	327	350
E1 (LO)	351	352	324	349

Further, we have focused on the ground state phonon density of states (PDOS) of WZ ZnS to clarify the contribution of the both elements (Zn and S) to the phonon properties of the compound. Figure 7 demonstrates the partial PDOS of WZ ZnS under zero pressure and temperature. As is apparent in Figure 7, the contribution of the Zn element to acoustic phonon modes is higher than S, whereas the opposite case is valid for the S element due to its dominant contribution to the optical modes of the compound.

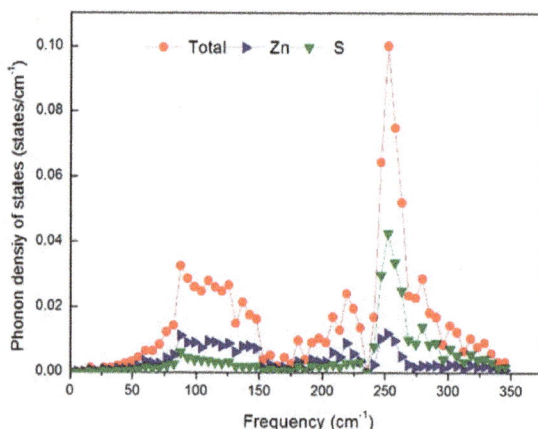

Figure 7. Partial and total phonon density of states of (PDOS) curve of WZ ZnS under zero temperature and pressure.

Figure 8 shows the phonon dispersion curves of WZ ZnS under different pressures. Each pressure value above 0 GPa shifts the dispersion curves of WZ ZnS up slightly to the higher frequency values, as is clear in Figure 8. This behavior strictly originates from the atoms of WZ ZnS which are getting closer to each other under pressure, since they sit in steeper potential wells. The effect of pressure also causes the same behavior in the total density of states curves of WZ ZnS as shown in Figure 9.

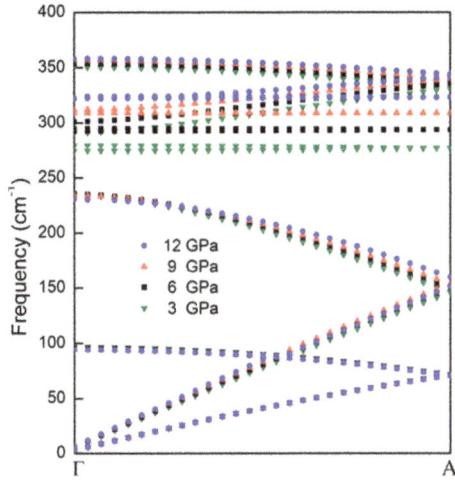

Figure 8. Phonon dispersion curves of WZ ZnS at zero temperature along the chosen Γ–A path under different pressures.

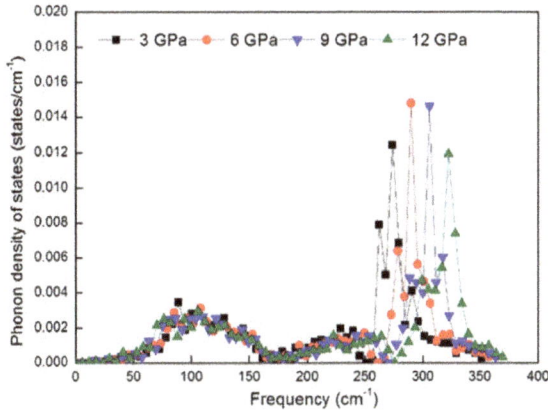

Figure 9. Total phonon density of states of (PDOS) curves of WZ ZnS under zero temperature with pressures 3 GPa, 6 GPa, 9 GPa, and 12 GPa.

Overall, the obtained results display a fair agreement with the experiments—in particular in elastic constants, bulk modulus, and phonon properties of WZ ZnS. Finally, the presented results for all considered quantities of WZ ZnS through this research are not only consistent with experiments, but also better than those of some published theoretical data.

4. Conclusions

In summary, we applied a mixed-type interatomic potential for the first time in conjunction with geometry optimization calculations to predict the high-pressure elastic, mechanical, and phonon properties of WZ ZnS. As our results demonstrate, the application of a mixed-type interatomic potential that is originally used for predicting the surface energies of zinc blende-type ZnS crystal structure well capture the elastic, mechanical and phonon features of WZ ZnS under pressure. From a quantitative evaluation, the obtained results of this work approximate the former experimental values—in particular for the elastic constants, bulk modulus, and phonon modes, and better than those of some other published data thanks to the sensitivity of the applied potential. The results of the present work may especially be helpful to future studies regarding the high-pressure elastic, mechanical, and other related properties of WZ ZnS.

Author Contributions: Melek Güler and Emre Güler conceived and designed the theoretical calculations; Melek Güler performed the all calculations; Melek Güler analyzed the all obtained data of work; both Melek Güler and Emre Güler wrote the paper.

Conflicts of Interest: The authors declare no conflicts of interest.

References

1. Bouhemadoua, A.; Allali, D.; Bin-Orman, S.; Muhammad Abud Al Safi, E.; Khenata, R.; Al-Douri, Y. Elastic and thermodynamic properties of the SiB2O4 (B=Mg, Zn and Cd) cubic spinels: An *ab initio* FP-LAPW study. *Mater. Sci. Semicond. Process.* **2015**, *38*, 192–202. [CrossRef]
2. Boudrifa, O.; Bouhemadou, A.; Guechi, N.; Bin-Omran, S.; Al-Douri, Y.; Khenata, R. First-principles prediction of the structural, elastic, thermodynamic, electronic and optical properties of Li4Sr3Ge2N6 quaternary nitride. *J. Alloy. Compd.* **2015**, *618*, 84–94. [CrossRef]
3. Boulechfar, R.; Meradji, H.; Chouahda, Z.; Ghemid, S.; Drablia, S.; Khenata, R. FP-LAPW investigation of the structural, electronic and thermodynamic properties of Al3Ta compound. *Int. J. Mod. Phys. B* **2015**, *29*, 1450244. [CrossRef]
4. Murtaza, G.; Gupta, S.K.; Seddik, T.; Khenata, R.; Alahmed, Z.A.; Ahmed, R.; Khachai, H.; Jha, P.K.; Bin Omran, S. Structural, electronic, optical and thermodynamic properties of cubic REGa3 (RE = Sc or Lu) compounds: Ab initio study. *J. Alloy. Compd.* **2014**, *597*, 36–44. [CrossRef]
5. Zeng, Z.; Christos, S.G.; Baskoutas, S.; Bester, G. Excitonic optical properties of wurtzite ZnS quantum dots under pressure. *J. Chem. Phys.* **2015**, *142*, 114305. [CrossRef] [PubMed]
6. Güler, E.; Güler, M. A Theoretical Investigation of the Effect of Pressure on the Structural, Elastic and Mechanical Properties of ZnS Crystals. *Braz. J. Phys.* **2015**, *45*, 296–301. [CrossRef]
7. Tan, J.J.; Li, Y.; Ji, G.F. High-Pressure Phase Transitions and Thermodynamic Behaviors of Cadmium Sulfide. *Acta Phys. Pol. A* **2011**, *120*, 501–506. [CrossRef]
8. Bilge, M.; Kart, S.Ö.; Kart, H.H.; Çağın, T. Mechanical and electronical properties of ZnS under pressure. *J. Ach. Mat. Man. Eng.* **2008**, *31*, 29–34.
9. Grünwald, M.; Zayak, A.; Neaton, J.B.; Geissler, P.L. Transferable pair potentials for CdS and ZnS crystals. *J. Chem. Phys.* **2012**, *136*, 234111. [CrossRef] [PubMed]
10. Rong, C.X.; Hu, C.E.; Yeng, Z.Y.; Cai, L.C. First-Principles Calculations for Elastic Properties of ZnS under Pressure. *Chin. Phys. Lett.* **2008**, *25*, 1064. [CrossRef]
11. Wright, K.; Gale, J.D. Interatomic potentials for the simulation of the zinc-blende and wurtzite forms of ZnS and CdS: Bulk structure, properties, and phase stability. *Phys. Rev. B* **2004**, *70*, 035211. [CrossRef]
12. Cheng, Y.C.; Jin, C.Q.; Gao, F.; Wu, X.L.; Zhong, W.; Li, S.H.; Chu, P.K. Raman scattering study of zinc blende and wurtzite ZnS. *J. App. Phys.* **2009**, *106*, 123505. [CrossRef]
13. Yu, Y.; Fang, D.; Zhao, G.D.; Zheng, X.L. Ab initio Calculation of the thermodynamic properties of of wurtzite ZnS: Performance of the LDA and GGA. *Chalcogenide Lett.* **2014**, *11*, 619–628.
14. Ferahtia, S.; Saib, S.; Bouarissa, N.; Benyettou, S. Structural parameters, elastic properties and piezoelectric constants of wurtzite ZnS and ZnSe under pressure. *Superlattices Microstruct.* **2014**, *67*, 88–96. [CrossRef]
15. Güler, M.; Güler, E. High pressure phase transition and elastic behavior of europium oxide. *J. Optoelectron. Adv. Mater.* **2014**, *16*, 1222–1227.
16. Ayala, A.P. Atomistic simulations of the pressure-induced phase transitions in BaF2 crystals. *J. Phys. Condens. Matter* **2001**, *13*, 11741. [CrossRef]
17. Güler, E.; Güler, M. Phase transition and elasticity of gallium arsenide under pressure. *Mater. Res.* **2014**, *17*, 1268–1272. [CrossRef]
18. Güler, E.; Güler, M. Elastic and mechanical properties of hexagonal diamond under pressure. *Appl. Phys. A* **2015**, *119*, 721–726. [CrossRef]

19. Dick, B.G.; Overhauser, A.W. Theory of the Dielectric Constants of Alkali Halide Crystals. *Phys. Rev.* **1958**, *112*, 90. [CrossRef]
20. Hamad, S.; Cristol, S.; Catlow, C.R.A. Surface Structures and Crystal Morphology of ZnS: Computational Study. *J. Phys. Chem. B* **2012**, *106*, 11002–11008. [CrossRef]
21. Gale, J.D. GULP: A computer program for the symmetry-adapted simulation of solids. *J. Chem. Soc. Faraday Trans.* **1997**, *93*, 629. [CrossRef]
22. Gale, J.D.; Rohl, A.L. The General Utility Lattice Program (GULP). *Mol. Simul.* **2003**, *29*, 291. [CrossRef]
23. Broyden, G.C. The Convergence of a Class of Double-rank Minimization Algorithms 1. General Considerations. *J. Inst. Math. Appl.* **1970**, *6*, 76. [CrossRef]
24. Fletcher, R. A new approach to variable metric algorithms. *Comput. J.* **1970**, *13*, 317. [CrossRef]
25. Goldfarb, D. A family of variable-metric methods derived by variational means. *Math. Comput.* **1970**, *24*, 23. [CrossRef]
26. Shanno, D.F. Conditioning of quasi-Newton methods for function minimization. *Math. Comput.* **1970**, *24*, 647. [CrossRef]
27. Monkhorst, H.J.; Pack, J.D. Special points for Brillouin-zone integrations. *Phys. Rev. B* **1976**, *13*, 5188. [CrossRef]
28. Güler, E.; Güler, M. High pressure elastic properties of wurtzite aluminum nitrate. *Chin. J. Phys.* **2014**, *52*, 1625–1635. [CrossRef]
29. Neumann, H. Lattice dynamics and related properties of $A^{I}B^{III}$ and $A^{II}B^{IV}$ compounds, I. Elastic constants. *Cryst. Res. Technol.* **2004**, *39*, 939–958. [CrossRef]
30. Hu, E.C.; Sun, L.L.; Zeng, Z.Y.; Chen, X.R. Pressure and Temperature Induced Phase Transition of ZnS from First-Principles Calculations. *Chin. Phys. Lett.* **2008**, *25*, 675–678.
31. Brafman, O.; Mitra, S.S. Raman Effect in Wurtzite- and Zinc-Blende-Type ZnS Single Crystals. *Phys. Rev.* **1968**, *171*, 931. [CrossRef]

crystals

MDPI

Article

Elastic, Mechanical and Phonon Behavior of Wurtzite Cadmium Sulfide under Pressure

Melek Güler * and Emre Güler

Department of Physics, Hitit University, Corum 19030, Turkey; eguler71@gmail.com
* Correspondence: mlkgnr@gmail.com; Tel.: +90-3642277000; Fax: +90-3642277005

Academic Editor: Matthias Weil
Received: 27 March 2017; Accepted: 1 June 2017; Published: 4 June 2017

Abstract: Cadmium sulfide is one of the cutting-edge materials of current optoelectronic technology. Although many theoretical works are presented the for pressure-dependent elastic and related properties of the zinc blende crystal structure of cadmium sulfide, there is still some scarcity for the elastic, mechanical, and phonon behavior of the wurtzitic phase of this important material under pressure. In contrast to former theoretical works and methods used in literature, we report for the first time the application of a recent shell model-based interatomic potential via geometry optimization computations. Elastic constants, elastic wave velocities, bulk, Young, and shear moduli, as well as the phonon behavior of wurtzite cadmium sulfide (w-CdS) were investigated from ground state to pressures up to 5 GPa. Calculated results of these elastic parameters for the ground state of w-CdS are approximately the same as in earlier experiments and better than published theoretical data. Our results for w-CdS under pressure are also reasonable with previous calculations, and similar pressure trends were found for the mentioned quantities of w-CdS.

Keywords: CdS; elastic; mechanical; phonon dispersion; wurtzite

1. Introduction

Recently, the computational predictions for materials has become a valuable and rapid way to resolve the unclear subjects of solid state physics. As well, calculating reasonable elastic and thermodynamic results for materials can substitute the impossible and expensive experiments and may provide deeper insights for the concerned materials [1–5].

The focus of the present work is a CdS compound from the II-VI semiconductor family (i.e., CdS, CdSe, and CdTe) which has widespread technological applications ranging from solar cells to light emitting diodes [6]. Further, under ambient conditions, CdS can crystallize in either zinc blende (ZB) crystal structure with space group F-43m or wurtzite (w) crystal structure with P6₃mc space group [7–11]. Moreover, as reported in recent experimental measurements [12], phase transition from the ZB phase to the wurtzite phase of CdS occurs between the pressure values 3.0 GPa and 4.3 GPa.

It is possible to find a number of works performed especially for the mechanical, elastic, thermodynamic, and other properties of ZB CdS compounds not only with experiments, but also with theoretical works due to its simple crystal structure [6]. In 2000, Benkabou et al. [7] surveyed the structural and elastic properties of several II-VI compounds (CdS, CdSe, ZnS, and ZnSe) in ZB phase with their determined Tersoff potential parameters. Later, in 2004, Wright and Gale [11] introduced their interatomic potentials for the structural and stability properties of ZB and w-CdS under zero Kelvin temperature and zero pressure (ground state conditions). Afterwards, in 2006, Deligoz et al. [8] performed norm-conserving pseudopotential density functional theory (DFT) calculations for ZB phase of CdX (X: S, Se, Te) compounds in their ground states. In 2011, Ouendadji and coworkers [6] computed the structural, electronic, and thermal characteristics of the ZB phase of CdS, CdSe, and CdTe compounds in their DFT studies by using full potential linearized augmented plane wave (FP-LAPW) through local density approximation (LDA) and the generalized gradient approximation (GGA) in their ground state. Grünwald et al. [9] also established transferable pair potentials for CdS and ZnS crystals to accurately describe the ground state features of ZB and w-CdS compounds in 2012. At the same time, Tan et al. [10] documented the effect of pressure and temperature on ZB and w-CdS structures in their plane-wave pseudopotential DFT study with LDA, including phase transition pressure, entropy, enthalpy, elasticity, free energy, and heat capacity. As is clear from these works, former investigations strictly focus on the ground state properties of the ZB CdS compound where w-CdS and its important physical properties under pressure are still lacking.

In this research, we therefore concentrate on the elastic, mechanical, and phonon properties of w-CdS under pressure. Contrary to the above applied methods and employed potentials of literature, we report for the first time the use of a recent shell model-type interatomic potential [13] to determine the mentioned high-pressure quantities of w-CdS with geometry optimization calculations. During our work, we considered the elastic constants, bulk modulus, shear modulus, and Young modulus, elastic wave velocities, mechanical stability conditions, and phonon properties of w-CdS under pressures between 0 GPa and 5 GPa at T = 0 K.

The next part of the manuscript (Section 2) gives a short outline for the geometry optimization and other details of present computations. Subsequently, Section 3 affords our results and earlier data on the calculated quantities of w-CdS with discussion. At the end, Section 4 summarizes the key results of the present survey in the conclusion of the paper.

2. Computational Methods

For crystals, geometry optimization is an effective and practical method utilized in both molecular dynamics (MD) and DFT computations to obtain a stable arrangement of periodic systems or molecules with rapid energy computations. The basic concept for geometry optimization deals with the repeated potential energy sampling in which energy shows a minimum and all acting forces on total atoms reach zero. Detailed information for geometry optimization method and further points can also be obtained from References [14–17].

All of the computations of this work were performed with the General Utility Lattice Program (GULP) 4.2 MD code. GULP can be used for wide-range property computations of periodic solids, surfaces, and clusters by applying an appropriate interatomic potential relevant to the demands of the research [14–17].

The most accurate and reliable results of computer simulations are strongly linked with the quality of the employed interatomic potentials during computations [18]. Besides, shell model-form interatomic potentials provide reasonable outcomes on both ground state and high-pressure features of fluorides, oxides, and other compounds [19,20]. Since the shell model and its methodologies are well-known [14–20], we present only a short explanation here. Most of the shell model potentials involve long-range Coulomb and short-range pairwise interactions, and their ionic polarization is treated by Dick and Overhauser [21]. Further, in the shell model, an atom is characterized by two discrete components: the core (signifies the core and nucleus electrons) and the shell (stands for the valence electrons). The core and the shell independently interact with other atoms and with each other. Therefore, the interaction potential used in this work was in the form of:

$$U_{ij} = \frac{q_i q_j}{r_{ij}} + A \exp\left(-\frac{r_{ij}}{\rho}\right) - \frac{C}{r_{ij}^6}$$

where the first term in the equation denotes the Coulomb interaction, the second term symbolizes the repulsive interaction of the overlapping electron clouds, and the third term holds for the van der Waals interaction. Additionally, A, ρ, and C are the particular Buckingham potential interaction parameters, where Coulomb interactions follow the Ewald summation method [22]. For more information and other conjectures, interested readers can see References [23–26].

A recent shell model-type interatomic potential [13] was employed in this work, which is originally derived from DFT calculations for the bulk properties of CdS, CdSe, PbS, and PbSe solid compounds as well as their mixed phases. To keep the original form of the applied potential, we also ignored the shell-shell interactions during the present research, as in Reference [13]. Further, Table 1 lists the present potential parameters employed in our calculations, and further details about the parameterization procedure of this potential can be also found in Reference [13]. The cell parameters of w-CdS were assigned as a = 4.13 Å and c = 6.63 Å, as seen in Figure 1 which is visualized with VESTA 3.0 [27]. Constant pressure optimization was applied to our work to avoid constraints for an efficient geometry optimization [14–17]. Additionally, cell geometries were optimized by Newton–Raphson procedure generated from the Hessian matrix. The Hessian matrix obtained from second derivatives of the energy was iteratively updated with the default BFGS algorithm [28–31] of GULP. After fixing the prerequisites for w-CdS crystal structure and initial optimization settings, we have performed various runs by ensuring the pressure values change from 0 GPa to 5 GPa with 0.5 GPa steps at zero Kelvin temperature.

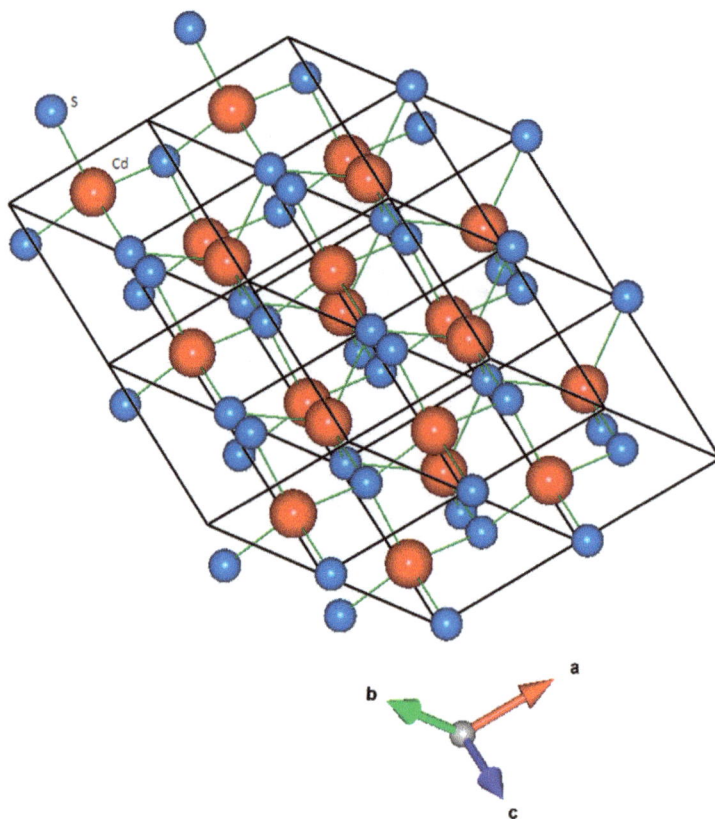

Figure 1. Crystal structure of wurtzitic CdS (w-CdS). Blue atoms show sulphur anions where red atoms represent cadmium cations along a, b, and c axes.

During our work, phonon and connected features of w-CdS were also considered after geometry optimizations as a function of external pressure through quasiharmonic approximation under zero temperature, as implemented in GULP. It is feasible to calculate the phonon density of states (PDOS) as well as phonon dispersions for a given material following the statement of a shrinkage factor via GULP phonon calculations. Besides, GULP defines the phonons by computing their values at special points in reciprocal space in the first Brillouin zone. To attain the Brillouin zone integration and determine the PDOS, we have employed a Monkhorts and Pack scheme [32] routine with $8 \times 8 \times 8$ *k-point* mesh.

Table 1. Shell model-type interatomic potential parameters taken from Reference [13]. The short-range interactions between shells (s) are ignored. The effective core (c) charges are assigned as 0.8e for Cd and -0.8e for S. A potential cut-off with radius 12 Å was set for short-range interactions.

Interaction	A (eV)	ρ (Å)	C (eV·Å^6)
Cd$_C$-S$_C$	1.26×10^9	0.107	53.5
S$_C$-S$_C$	4.68×10^3	0.374	120

3. Results and Discussion

Figure 2 represents the density behavior of w-CdS under pressure. The density of many materials displays straight increments under pressure due to the volume reduction of the related crystal. This fact is also valid for w-CdS under pressure, as seen in Figure 2. The lowest density value of w-CdS is 4.89 g/cm^3 for 0 GPa, and the highest density value at 5 GPa is 5.2 g/cm^3 at zero temperature. The presently obtained ground state (T = 0 K and P = 0 GPa) value of density with 4.89 g/cm^3 of w-CdS is comparable to the room temperature experimental density value of 4.82 g/cm^3 [33].

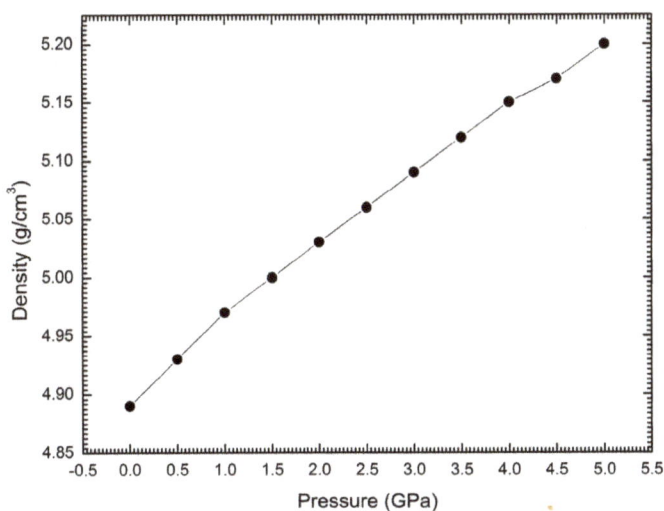

Figure 2. Density behavior of w-CdS under pressure.

After optimizing the structure of a given material, it is then possible to compute different physical features with GULP. These calculations comprise the elastic constants, bulk modulus, and other mechanical quantities (shear modulus and Young moduli, elastic wave velocities, etc.) of the regarding material. For instance, the presently calculated elastic constants show the second derivatives of the energy density with respect to strain, and details about other remaining property calculations can be also found within Reference [16].

Elastic constants deliver clear perceptions about the mechanical and other associated properties of materials. Though elastic constants obtained from total energy computations belong to single crystal values, it is crucial to get accurate polycrystalline elastic constants of materials because many technically important materials exist in polycrystalline form [34]. For this reason, Voigt–Reuss–Hill [35–37] values were considered during this work.

For wurtzite crystals, five well-known elastic constants exist, which are specified as C_{11}, C_{12}, C_{13}, C_{33}, and C_{44} [38].

Figure 3 indicates our findings for C_{11}, C_{12}, C_{13}, C_{33}, and C_{44} constants for w-CdS under pressures between 0 GPa and 5 GPa. Among them C_{11} and C_{33} represent the longitudinal elastic character where elastic wave propagation occurs easily under pressure. This easiness causes the increments of C_{11} and C_{33} under pressure. Surprisingly, the magnitudes of C_{11} and C_{33} are similar to each other in the ground state, and the gap between them becomes greater as the pressure increases. Unlike C_{11} and C_{33}, C_{44} constant (characterizing the shear elastic response to retarded wave propagation) has a sluggish decrement. Figure 3 also shows that the calculated elastic constants are in the range of $C_{33} > C_{11} > C_{12} > C_{13} > C_{44}$. Both the range of the elastic constants and the slight decrement of the C_{44} constant under pressure mimic the DFT findings of Tan et al. [10]. However, our results for the ground state parameters of w-CdS are obviously satisfactory compared to those of by Tan et al. [10] as well as Wright and Gale [11], and much closer to the experimental results of Bolef et al. [39] as listed in Table 2.

According to stability, the proverbial Born mechanical stability criterion for hexagonal crystals is also valid for wurtzite crystals, and must fulfill [38]:

$$C_{44} > 0, C_{11} > |C_{12}| \text{ and } (C_{11} + 2\,C_{12})\,C_{33} > 2C_{13}^2$$

Presently calculated results of elastic constants of w-CdS obey the mechanical stability criterion (Figure 3 and Table 2), which consequently indicates that the w-CdS is mechanically stable in its ground state.

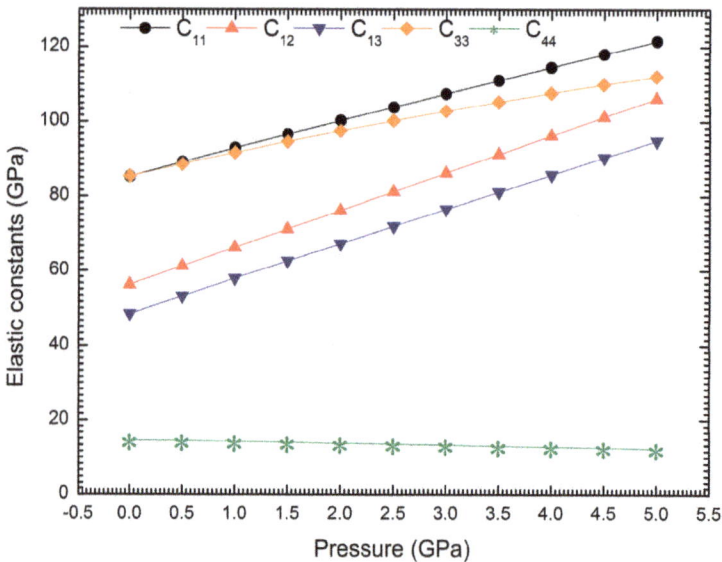

Figure 3. Elastic constants of w-CdS under pressure.

Bulk modulus (B) is an essential elastic constant connected to the bonding strength and is used as a primary parameter for the calculation of a material's hardness. Shear modulus (G) is the measure of the resistivity of a material after applying a shearing force. Furthermore, Young modulus (E) also defines the amount of a material's resistance to uniaxial tensions. These three distinct moduli (B, G, and *E*) are other valuable parameters for classifying the mechanical properties of materials. Figure 4 shows the bulk modulus, Young modulus, and shear modulus (*B*, *E*, and *G*) of w-CdS under pressure. From the common physical expression of bulk modulus $\left(B = \Delta \frac{P}{\Delta} V \right)$, it is not difficult to predict a raise for *B* due to its direct proportion to pressure. So, the bulk modulus of w-CdS represents a clear increment with pressure. Conversely, *G* and *E* moduli have insignificant decrements under pressure, again similar to the DFT findings of Tan et al. [10]. Moreover, Table 2 also lists numerical comparisons for *B*, *G*, and *E* moduli of current and earlier data of w-CdS under zero temperature and pressure. Our results for *B*, *G*, and *E* moduli agree with both experiments and other DFT results, as seen in Table 2.

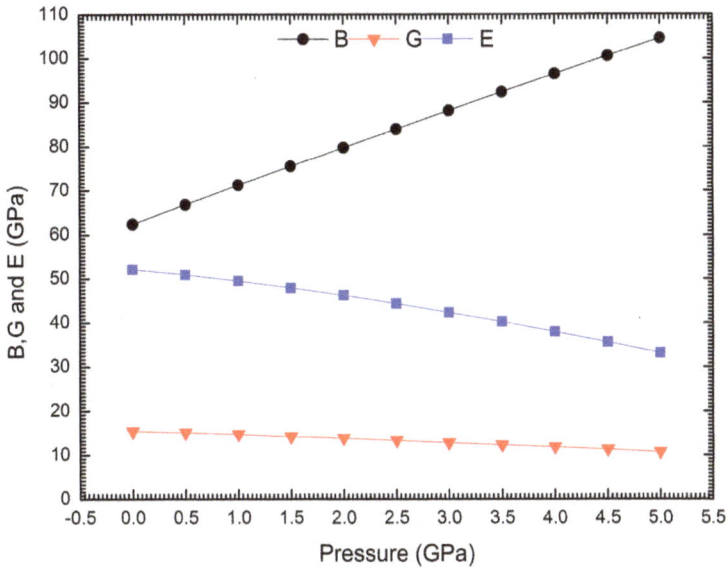

Figure 4. Elastic moduli (*B*, *G*, and *E*) behavior of w-CdS versus pressure.

Ductile and brittle responses of materials represent two antithetical mechanical characteristics of solids when they exposed to external stress. Since these adjectives (ductility and brittleness) are important for the production of desired materials, we also checked the ductile (brittle) behavior of w-CdS under pressure. Usually, brittle materials display a considerable resistance to the deformation before fracture, whereas ductile materials can be easily deformed. In addition, Pugh ratio evaluation [40] is one of the prevalent routines in the literature which conveys a decisive limit for ductile (brittle) performance of materials. As stated by Pugh, a material can be ductile if its *G*/*B* ratio is smaller than 0.5; otherwise, it can be brittle. Our careful assessment shows that *G*/*B* values decrease from 0.24 (P = 0 GPa) to 0.10 (P = 5 GPa) at zero temperature for w-CdS as in Figure 5. So, w-CdS behaves in a ductile manner for the entire pressure range.

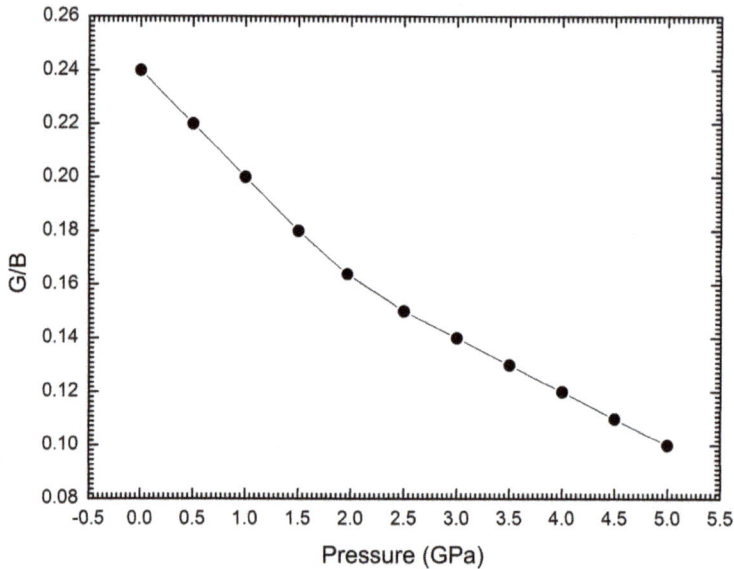

Figure 5. G/B ratio of w-CdS against pressure.

Longitudinal and shear elastic waves may arise in solids at low temperatures due to vibrational excitations originating from the acoustic modes [20]. Thus, V_L signifies the longitudinal elastic wave velocity, where V_S stands for the shear wave velocity. Figure 6 represents the pressure behavior of V_L and V_S of w-CdS pressure at T = 0 K. As is clear in Figure 6, V_L has a significant increment compared to V_S, and this is the most common case for materials because of the facts explained above. Obtained data of this work for both V_L and V_S are again reasonable when compared to previous experiments (See Table 2).

Table 2. Comparing the present results with former experimental and theoretical data for the calculated parameters of w-CdS at zero pressure and temperature.

Parameter	Exp [33]	Present	Ref [9]	Ref [10]	Ref [11]	Ref [41]
C_{11} (GPa)	84.3	85.2	107.3	93.9	102.8	80.5
C_{12} (GPa)	52.1	56.2	35.8	57.6	45.4	45.0
C_{13}(GPa)	46.3	48.4	15.9	50.1	47.5	37.1
C_{33} (GPa)	93.9	85.3	144.3	105.2	113.3	87.0
C_{44} (GPa)	14.8	14.5	20.5	15.8	32.4	15.2
B (GPa)	62.7	62.4	54.0	68.9	66.4	54.0
E (GPa)	48.1	52.0		51.0		
G (GPa)	-	15.4		18.5		
V_S (km/s)	1.84	1.77				
V_L (km/s)	4.24	4.11				

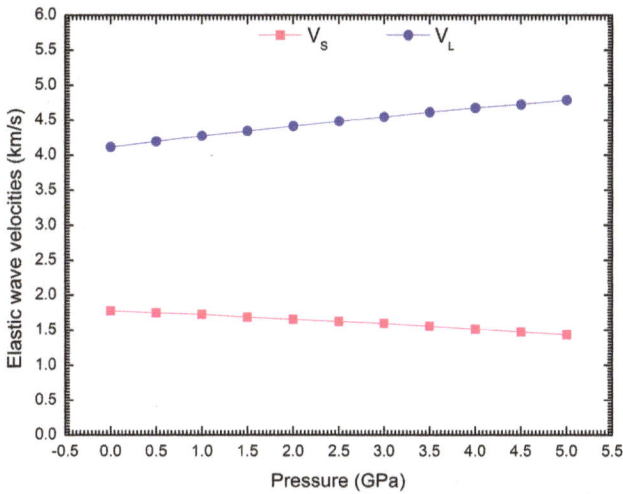

Figure 6. Behavior of longitudinal (V_L) and shear wave (V_S) velocities of w-CdS under pressure.

The success of the present potential on the ground state phonon dispersion properties of w-CdS have already been quantitatively compared in its original reference [13] with experimental results. Additionally, we would like to present the missing ground state phonon density of states (PDOS) of w-CdS to explain the contribution of both elements S and Cd to the phonon properties of the compound. Figure 7 displays the partial and total PDOS of w-CdS in the ground state. In Figure 7, the phonon density of states appear with four well-separated regions corresponding to the longitudinal and transverse acoustic modes (LA and TA) and longitudinal and transverse optic modes (LO and TO) of w-CdS. Besides, the contribution of the Cd element to acoustic phonon modes is higher than S, whereas the opposite case is valid for the S element due to its dominant contribution to optical modes. There is also a clear gap between the frequencies 150 cm^{-1} and 250 cm^{-1} originating from the mass differences of Cd and S elements of w-CdS.

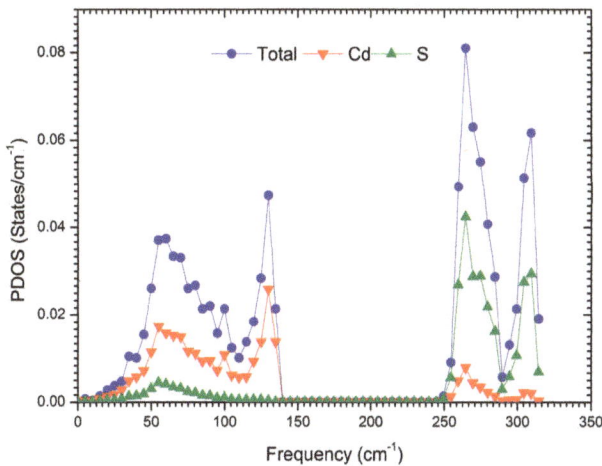

Figure 7. Partial and total phonon density of states of (PDOS) curve of w-CdS under zero temperature and pressure.

On the other hand, Figure 8 shows the phonon dispersion of w-CdS along the chosen Γ-A path (as same as with original Ref [13]) in reciprocal space for pressures 0 GPa, 1 GPa, 3 GPa, and 5 GPa. As is evident in Figure 8, each pressure value above 0 GPa slightly shifts the phonon dispersion curves of w-CdS to higher frequency values of due to atoms which move towards to each other and sit in steeper potential wells under pressure. The corresponding PDOS curves of applied pressures are also given in Figure 9. The increasing pressure also increases the PDOS peaks of w-CdS for each pressure, and under pressure, the gap between acoustic and optical modes shifts to higher frequencies from 150 cm^{-1} to 275 cm^{-1}.

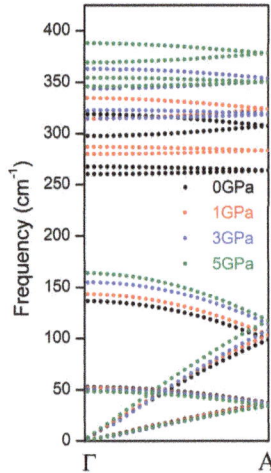

Figure 8. Phonon dispersion curve of w-CdS at zero temperature under pressures 0 GPa, 1 GPa, 3 GPa, and 5 GPa.

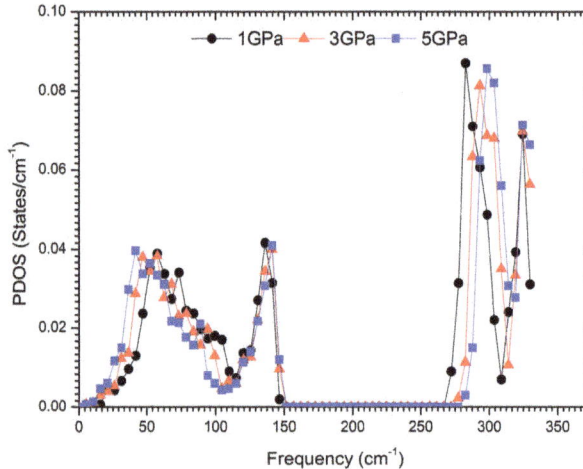

Figure 9. Phonon density of states of (PDOS) curve of w-CdS at zero temperature for 1 GPa, 3 GPa, and 5 GPa.

Overall, our results for this research demonstrate a fair accordance with the experiments—especially for elastic constants, bulk modulus, elastic wave velocities, and phonon properties of w-CdS in its ground state. Finally, the presented results for all calculated parameters of w-CdS through this work are both consistent with experiments and better than those of some published theoretical data.

4. Conclusions

In summary, we applied a recent shell model-type interatomic potential for the first time with geometry optimization calculations to study both ground state and pressure-dependent elastic, mechanical, and phononic aspects of w-CdS. As the present results prove, the application of this potential which is originally employed for computing the ground state bulk properties of w-CdS also successfully captures the investigated properties under pressure. Particularly, present results for the ground state of w-CdS are about former experiments for the elastic constants, bulk modulus, elastic wave velocities, and phonon characteristics and better than those of other published theoretical data. Moreover, the effect of pressure on w-CdS was also presented and reasonable results were obtained for several properties of w-CdS after benchmarking the existing literature. Bulk modulus, shear modulus, and other longitudinal wave-related elastic and mechanical quantities show clear increments under pressure, where the shear wave connected parameters display sluggish decrements due to the nature of longitudinal and shear elastic waves propagation. w-CdS exhibits ductile character in its ground state and even under pressure. We hope that our results add value to the forthcoming researches about w-CdS under pressure.

Author Contributions: Melek Güler and Emre Güler conceived and designed the theoretical calculations; Melek Güler performed the all calculations; Melek Güler analyzed the all obtained data of work; both Melek Güler and Emre Gülerwrote the paper.

Conflicts of Interest: The authors declare no conflict of interest.

References

1. Ullah, N.; Ullah, H.; Murtaza, G.; Khenata, R.; Ali, S. Structural phase transition and optoelectronic properties of ZnS under pressure. *J. Opt. Adv. Mater.* **2015**, *17*, 1272.
2. Bouhemadou, A.; Haddadi, K.; Khenata, R.; Rached, D.; Bin-Omran, S. Structural, elastic and thermodynamic properties under pressure and temperature effects of MgIn2S4 and CdIn2S4. *Phys. B Condens. Matter* **2012**, *407*, 2295. [CrossRef]
3. Seddik, T.; Semari, F.; Khenata, R.; Bouhemadou, A.; Amrani, B. High pressure phase transition and elastic properties of Lutetium chalcogenides. *Phys. B Condens. Matter* **2010**, *405*, 394. [CrossRef]
4. Duan, D.; Liu, Y.; Ma, Y.; Liu, Z.; Cui, T.; Liu, B.; Zou, G. Ab initio studies of solid bromine under high pressure. *Phys. Rev. B* **2007**, *76*, 104113. [CrossRef]
5. Duan, D.; Huang, X.; Tian, F.; Li, D.; Yu, H.; Liu, Y.; Ma, Y.; Liu, B.; Cui, T. Pressure-induced decomposition of solid hydrogen sulfide. *Phys. Rev. B* **2015**, *91*, 180502(R). [CrossRef]
6. Ouendadji, S.; Ghemid, S.; Meradji, H.; El Haj Hassan, F. Theoretical study of structural, electronic, and thermal properties of CdS, CdSe and CdTe compounds. *Comput. Mater. Sci.* **2011**, *50*, 1460. [CrossRef]
7. Benkabou, F.; Aourag, H.; Certier, M. Atomistic study of zinc-blende CdS, CdSe, ZnS, and ZnSe from molecular dynamics. *Mater. Chem. Phys.* **2000**, *66*, 10–16. [CrossRef]
8. Deligoz, E.; Colakoglu, K.; Ciftci, Y. Elastic, electronic, and lattice dynamical properties of CdS, CdSe, and CdTe. *Phys. B Condens. Matter* **2006**, *373*, 124–130. [CrossRef]
9. Grünwald, M.; Zayak, A.; Neaton, J.B.; Geissler, P.L.; Rabani, E. Transferable pair potentials for CdS and ZnS crystals. *J. Chem. Phys.* **2012**, *136*, 234111. [CrossRef] [PubMed]
10. Tan, J.J.; Li, Y.; Ji, G.F. High-Pressure Phase Transitions and Thermodynamic Behaviors of Cadmium Sulfide. *Acta Phys. Pol. A* **2011**, *120*, 501–506. [CrossRef]
11. Wright, K.; Gale, J.D. Interatomic potentials for the simulation of the zinc-blende and wurtzite forms of ZnS and CdS: Bulk structure, properties, and phase stability. *Phys. Rev. B* **2004**, *70*, 035211. [CrossRef]
12. Li, Y.; Zhang, X.; Li, H.; Li, X.; Lin, C.; Xiao, W.; Liu, J. High pressure-induced phase transitions in CdS up to 1 Mbar. *J. Appl. Phys.* **2013**, *113*, 083509. [CrossRef]

13. Fan, Z.; Koster, R.S.; Wang, S.; Fang, C.; Yalcin, A.O.; Tichelaar, F.D.; Zandbergen, H.W.; van Huis, M.A.; Vlugt, T.J.H. A transferable force field for CdS-CdSe-PbS-PbSe solid systems. *J. Chem. Phys.* **2014**, *141*, 244503. [CrossRef] [PubMed]

14. Gale, J.D. Empirical potential derivation for ionic materials. *Philos. Mag. B* **1996**, *73*, 3–19. [CrossRef]

15. Gale, J.D. GULP: A computer program for the symmetry-adapted simulation of solids. *J. Chem. Soc. Faraday* **1997**, *93*, 629. [CrossRef]

16. Gale, J.D.; Rohl, A.L. The General Utility Lattice Program (GULP). *Mol. Simulat.* **2003**, *29*, 291. [CrossRef]

17. Gale, J.D. GULP: Capabilities and prospects. *Z. Kristallogr.* **2005**, *220*, 552. [CrossRef]

18. Walsh, A.; Sokol, A.A.; Catlow, C.R.A. *Computational Approaches to Energy Materials*, 1st ed.; John Wiley & Sons, Ltd.: West Sussex, UK, 2013; p. 101.

19. Valerio, M.E.G.; Jackson, R.A.; de Lima, J.F. Derivation of potentials for the rare-earth fluorides, and the calculation of lattice and intrinsic defect properties. *J. Phys. Condens. Matter* **2000**, *12*, 7727. [CrossRef]

20. Güler, E.; Güler, M. High pressure elastic properties of wurtzite aluminum nitrate. *Chin. J. Phys.* **2014**, *52*, 1625–1635.

21. Dick, B.G.; Overhauser, A.W. Theory of the Dielectric Constants of Alkali Halide Crystals. *Phys. Rev.* **1958**, *112*, 90. [CrossRef]

22. Ewald, P.P. Die Berechnung optischer und elektrostatischer Gitterpotentiale. *Ann. Phys.* **1921**, *64*, 253. [CrossRef]

23. Archer, T.D.; Birse, S.E.A.; Dove, M.T.; Redfern, S.; Gale, J.D.; Cygan, R.T. An interatomic potential model for carbonates allowing for polarization effects. *Phys. Chem. Miner.* **2003**, *30*, 416. [CrossRef]

24. Ayala, A.P. Atomistic simulations of the pressure-induced phase transitions in BaF2 crystals. *J. Phys. Condens. Matter* **2001**, *13*, 11741. [CrossRef]

25. Chisholm, J.A.; Lewis, D.W.; Bristowe, P.D. Classical simulations of the properties of group-III nitrides. *J. Phys. Condens. Matter* **1999**, *11*, L235–L239. [CrossRef]

26. Kilo, M.; Jackson, R.A.; Borchardt, G. Computer modelling of ion migration in zirconia. *Phil. Mag.* **2003**, *83*, 3309. [CrossRef]

27. Momma, K.; Izumi, F. VESTA 3 for three-dimensional visualization of crystal, volumetric and morphology data. *J. Appl. Crystallogr.* **2011**, *44*, 1272–1276. [CrossRef]

28. Broyden, G.C. The Convergence of a Class of Double-rank Minimization Algorithms 1. General Considerations. *J. Inst. Math. Appl.* **1970**, *6*, 76. [CrossRef]

29. Fletcher, R. A new approach to variable metric algorithms. *Comput. J.* **1970**, *13*, 317. [CrossRef]

30. Goldfarb, D. A family of variable-metric methods derived by variational means. *Math. Comput.* **1970**, *24*, 23. [CrossRef]

31. Shanno, D.F. Conditioning of quasi-Newton methods for function minimization. *Math. Comput.* **1970**, *24*, 647. [CrossRef]

32. Monkhorst, H.J.; Pack, J.D. Special points for Brillouin-zone integrations. *Phys. Rev. B* **1976**, *13*, 5188. [CrossRef]

33. Adachi, S. *Handbook on Physical Properties of Semiconductors*; Kluwer: Boston, MA, USA, 2004; Volume 1.

34. Hirsekorn, S. Elastic properties of polycrystals. *Textures Microstruct.* **1990**, *12*, 1–14. [CrossRef]

35. Voigt, W. *Lehrbuch der Kristallphysik*; B.G. Teubner: Leipzig, Germany, 1928.

36. Reuss, A. Berechnung der Fließgrenze von Mischkristallen auf Grund der Plastizitätsbedingung für Einkristalle. *Angew. Z. Math. Mech.* **1929**, *9*, 55. [CrossRef]

37. Hill, R. The Elastic Behaviour of a Crystalline Aggregate. *Proc. Phys. Soc. A* **1952**, *65*, 349. [CrossRef]

38. Born, M.; Huang, K. *Dynamical Theory of Crystal Lattices*; Clarendon: Oxford, UK, 1956.

39. Bolef, D.I.; Melamed, N.T.; Menes, M. Elastic constants of hexagonal cadmium sulfide. *J. Phys. Chem. Solids* **1960**, *17*, 143. [CrossRef]

40. Pugh, S.F. Relations between the elastic moduli and the plastic properties of polycrystalline pure metals. *Philos. Mag.* **1954**, *45*, 823. [CrossRef]

41. Guo, X.J.; Xu, B.; Liu, Z.Y.; Yu, D.L.; He, J.L.; Guo, L.C. Theoretical Hardness of Wurtzite-Structured Semiconductors. *Chin. Phys. Lett.* **2008**, *25*, 2158.

crystals

MDPI

Article

Morphological and Crystallographic Characterization of Primary Zinc-Rich Crystals in a Ternary Sn-Zn-Bi Alloy under a High Magnetic Field

Lei Li [1,2], Chunyan Ban [1,2,*], Ruixue Zhang [1,2], Haitao Zhang [1,2], Minghui Cai [2], Yubo Zuo [1,2], Qingfeng Zhu [1,2], Xiangjie Wang [1,2] and Jianzhong Cui [1,2]

[1]   Key Laboratory of Electromagnetic Processing of Materials, Ministry of Education, Northeastern University, Shenyang 110819, China; lilei@epm.neu.edu.cn (L.L.); zrx532@163.com (R.Z.); haitao_zhang@epm.neu.edu.cn (H.Z); zuoyubo@epm.neu.edu.cn (Y.Z.); zhuqingfeng@epm.neu.edu.cn (Q.Z.); wangxj@epm.neu.edu.cn (X.W.); jzcui@epm.neu.edu.cn (J.C.)
[2]   School of Materials Science and Engineering, Northeastern University, Shenyang 110819, China; cmhing@126.com
*   Correspondence: bancy@epm.neu.edu.cn; Tel.: +86-24-8368-1895

Academic Editor: Matthias Weil
Received: 3 June 2017; Accepted: 3 July 2017; Published: 6 July 2017

Abstract: Due to the unique capacity for structural control, high magnetic fields (HMFs) have been widely applied to the solidification process of alloys. In zinc-based alloys, the primary zinc-rich crystals can be dendritic or needle-like in two dimensions. For the dendritic crystals, their growth pattern and orientation behaviors under HMFs have been investigated. However, the three-dimensional crystallographic growth pattern and the orientation behaviors of the needle-like primary zinc-rich crystals under a high magnetic field have not been studied. In this work, a ternary Sn-Zn-Bi alloy was solidified under different HMFs. The above-mentioned two aspects of the needle-like primary zinc-rich crystals were characterized using the Electron Backscattered Diffraction (EBSD) technique. The results show that the primary zinc-rich crystals are characterized by the plate-shaped faceted growth in three dimensions. They grow in the following manner: spreading rapidly in the {0001} basal plane with a gradual decrease in thickness at the edges. The application of HMFs has no effect on the growth form of the primary zinc-rich crystals, but induces their vertical alignment. Crystallographic analysis indicates that the vertically aligned primary zinc-rich crystals orient preferentially with the *c*-axis perpendicular to the direction of the magnetic field.

Keywords: high magnetic field; solidification; zinc-rich crystal; characterization; crystallography; EBSD

1. Introduction

Since the last century, great attention has been devoted to the application of a magnetic field to the solidification process of alloys due to its unique capacity for structural control [1–23]. Before the development of superconducting magnets, a conventional magnetic field (usually smaller than 2 T) was the most extensively used in research and practice. Owing to the low intensities, conventional magnetic fields mainly affect the structures of alloys through suppressing or driving the fluids by the induced Lorentz force [1–5]. With the recent development of the superconducting magnet, high magnetic fields (HMFs) (usually larger than 2 T) have become readily available and have been widely introduced in the solidification process of alloys. When compared to the conventional magnetic fields, the HMFs significantly enhance the Lorentz force and the magnetization force (usually negligible in conventional magnetic fields), thereby imposing more abundant effects on the structures of the alloys [6–23]. It has been found that the HMFs could orient and align the structures [13–17], increase the phase transformation temperature [19], enhance the magnetic coercivities [16], suppress the diffusion of solute elements [21], modify the orientation relationship between the eutectics [22], and change the solid-liquid interface morphologies [23].

Hexagonal close-packed zinc is characterized by large solid-liquid interfacial energy anisotropy [24,25] and magnetocrystalline anisotropy [13,14]. Therefore, it is of fundamental interest to study its growth under HMFs. In other work, it has been shown that the primary zinc-rich crystals in Zn-Al [13] and Zn-Sn [14,26] alloys have a dendritic form (preferentially grow along $\langle 10\bar{1}0 \rangle$ and/or <0001>) and can be preferentially aligned and oriented by HMFs. According to some recent work, the primary zinc-rich crystals in ternary Sn-Zn-Bi solder alloys usually exhibit a needle-like morphology in two dimensions [27]. However, their crystallographic growth pattern in three dimensions has not been addressed. The crystallographic effects of HMFs on such crystals are also not clear. Based on this context, a ternary Sn-12Zn-6Bi alloy (nominal composition: wt %) was solidified under various HMFs in this work. The three-dimensional growth form, the alignments, and the orientations of the primary zinc-rich crystals in the alloy were characterized using Electron Backscattered Diffraction (EBSD). The affecting mechanism of the HMFs on the alignments and orientations of the primary zinc-rich crystals was discussed briefly.

2. Results and Discussion

Figure 1 shows the differential scanning calorimetry (DSC) curve of the Sn-12Zn-6Bi alloy upon cooling, where two exothermic peaks are detected. This is similar to the case of the Sn-8Zn-6Bi alloy [27], according to which the reaction orders should be as follows: L → L + primary Zn → L + primary Zn + secondary Sn → primary Zn + secondary Sn + eutectic (Sn + Zn). Furthermore, as the solubility of Bi in Sn decreases with the drop in the temperature, Bi crystals will be precipitated from the Sn matrix. The small peak in the curve corresponds to the crystallization of the primary Zn in the melt.

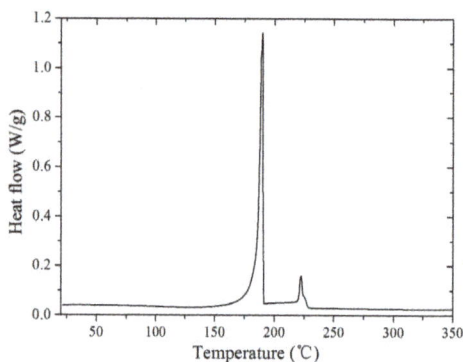

Figure 1. DSC curve of the ternary Sn-12Zn-6Bi alloy upon cooling.

Figure 2 shows the longitudinal macrostructures of the Sn-Zn-Bi specimens under various HMFs. As can be observed, the large dark primary zinc-rich crystals reveal a typical needle-like form. Obviously, the alignments of these primary crystals are heavily affected by the application of HMFs. Without and with a 0.6 T HMF, they are randomly aligned (Figure 2a,b). When increasing the HMF to 1.5 and 3 T, some crystals in the central regions of the specimens tend to align vertically, i.e., with the longer axis parallel to HMF direction *B* (Figure 2c,d). When the HMF is increased to 5 T, almost all of the crystals in the central regions align vertically. However, a further increase of the HMF to 12 T does not further enhance the alignment tendency. Moreover, it should be noted that the needle-like zinc-rich crystals in the peripheral regions are less affected by the application of the HMFs, i.e., they align randomly in all cases.

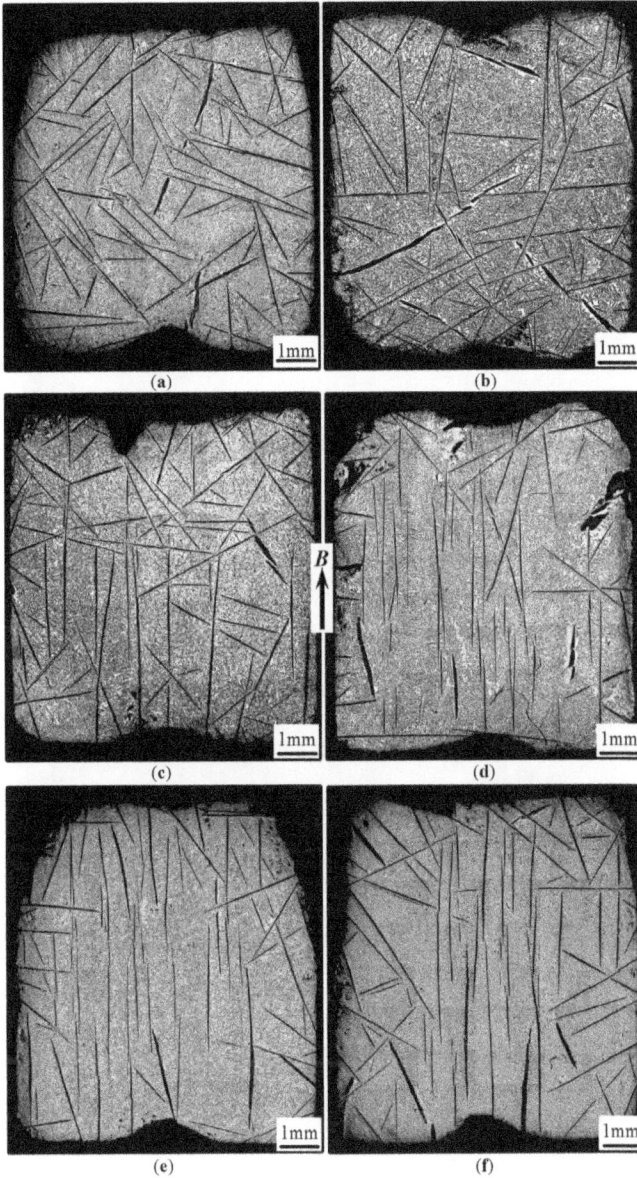

Figure 2. Longitudinal macrostructures of the Zn-Sn-Bi alloy under various magnetic fields: (**a**) 0 T; (**b**) 0.6 T; (**c**) 1.5 T; (**d**) 3 T; (**e**) 5 T; (**f**) 12 T. The arrow indicates the direction of *B*.

For more details, Figure 3 shows the microstructures in the central regions of the specimens under various HMFs, in which the regular alignments of the needle-like zinc-rich crystals can be more clearly observed. Other than this, it can also be seen that many fine dark-gray fibers are distributed randomly in the white-gray matrix surrounding the large primary zinc-rich crystals. Some small medium-gray globular crystals also exist and attach to each fiber to form a string-of-pearls-like morphology, as shown in the magnified insert in Figure 3a. A microstructural analysis indicates that the dark-gray fibers, the white-gray matrix, and the medium-gray globes are the eutectic zinc-rich phase, β-Sn matrix, and Bi crystals, respectively. Here, it should be mentioned that the various contrasts and morphologies of the eutectics are related to the polishing time: as the polishing solution is corrosive, different polishing times will result in different appearances (e.g., Figure 3b,c correspond to a longer polishing time so that many globular Bi crystals are revealed). However, it is certain that the random alignments of the fiber-like eutectic zinc-rich phase are unaffected by the HMF.

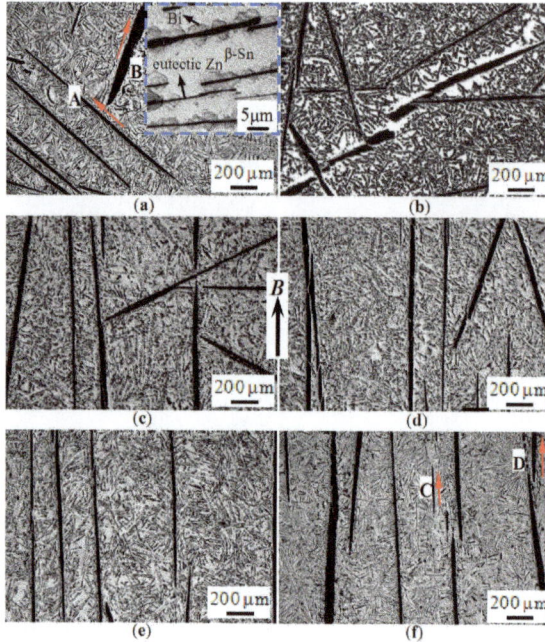

Figure 3. Microstructures corresponding to the central regions of the macrostructures in Figure 2: (a) 0, (b) 0.6, (c) 1.5, (d) 3, (e) 5, and (f) 8.8 T, respectively. The insert in (a) shows the magnified microstructure consisting of eutectic zinc-rich phase, β-Sn matrix, and Bi crystals, respectively. A, B, C, and D denote the crystals that will be further analyzed in Figure 6. The arrow indicates the direction of *B*.

To further understand the effects of the HMFs, Figure 4 shows the transverse macrostructures of the specimens. As can be observed, the large primary zinc-rich crystals still reveal a needle-like form, but align randomly in all cases. However, the unchanged needle-like form indicates that the large zinc-rich crystals may be plate-shaped in three dimensions. To identify this, Figure 5a,b show the microstructures of the cubes cut from the 0 and 12 T specimens, respectively. It can be seen that the primary zinc-rich crystals in surfaces 1, 2, and 3 connect perfectly at the edges of both of the cubes, confirming a plate shape in three dimensions, irrespective of whether the HMF is applied. Nevertheless, the needle-shaped form (with sharp ends) in two dimensions suggests that the edges of these plates should have a tapered transition in thickness. Furthermore, the primary zinc-rich plates exhibit some faceted growth character, as can be seen from the relatively smooth interface traces. A crystallographic interface calculation indicates that the large surfaces of the plates correspond to the {0001} basal plane. This result can also be indirectly proven by the <0001> pole figures in Figure 6a,b, corresponding to the crystals A & B in Figure 3a and C & D in Figure 3f, respectively. The projection lines OA, OB, OC, and OD are approximately perpendicular to the interface traces of the crystals A, B, C, and D in Figure 3 (see the arrows denoting the extension directions of the interfaces), respectively.

Furthermore, it should be mentioned that crystal B in Figure 3a is coarser than the others. This is because that it approximately exposes the {0001} basal plane in the longitudinal section, as evidenced by pole A near the center point O in Figure 6a. From the analysis above, it can be concluded that the plate-shaped primary zinc-rich crystals should grow in three dimensions, as follows: spreading rapidly in the {0001} basal plane and then decreasing gradually in thickness at the edges. Figure 7 schematically shows the three-dimensional morphology of a primary zinc-rich crystal, in conjunction with the hexagonal unit cells, denoting its orientations relative to the observation plane. It should be mentioned here that the growth manner of the primary zinc-rich crystals in the present alloy is quite different from that of the crystals in Zn-Sn and Zn-Al alloys, which, as mentioned previously, grow along the <$10\bar{1}0$> and/or <0001> directions to form complex dendritic morphologies [13,14,26]. Such a growth manner may be related to the types and amounts of alloying elements, which affect the growth kinetics of zinc-rich crystals. This is open to future research.

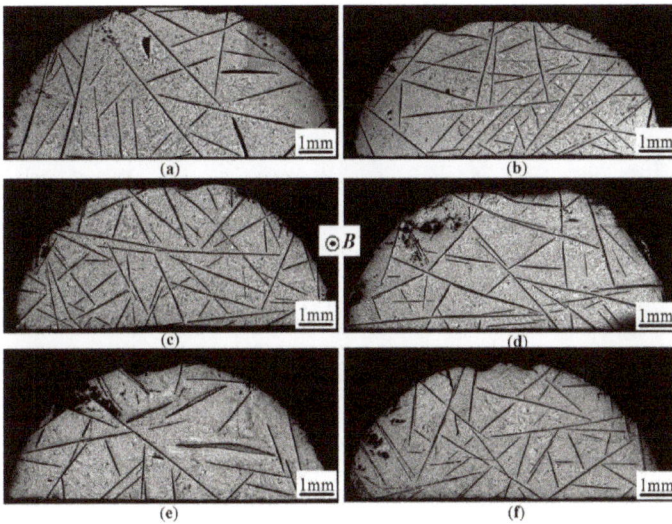

Figure 4. Transverse macrostructures of the specimens under various HMFs: (**a**) 0 T; (**b**) 0.6 T; (**c**) 1.5 T; (**d**) 3 T; (**e**) 5 T; (**f**) 12 T.

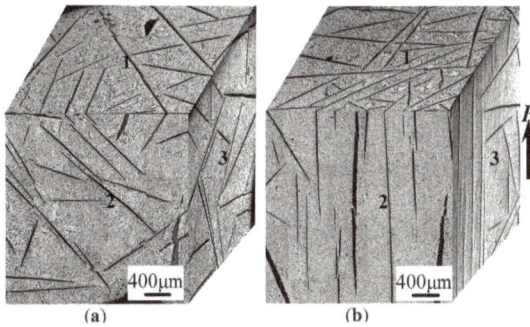

Figure 5. Microstructures of the cubes cut from the 0 and 12 T specimens. 1, 2, and 3 indicate the different surfaces of the cubes. The arrow indicates the direction of *B*.

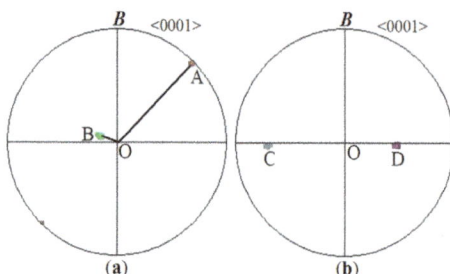

Figure 6. <0001> pole figures corresponding to the primary zinc-rich crystals (**a**) A & B in Figure 3a and (**b**) C & D in Figure 3f, respectively.

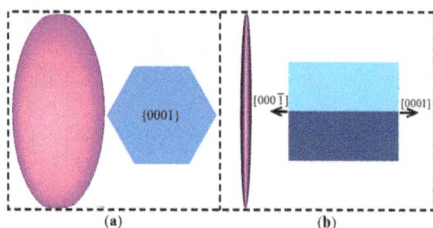

Figure 7. Schematic morphology of a primary zinc-rich crystal in three dimensions: (**a**) front and (**b**) side views. The hexagonal unit cells denote its orientations relative to the observation plane.

It has been optically observed that the primary zinc-rich crystals in the Sn-Zn-Bi alloy are highly aligned owing to the application of the HMFs. To discover more structural transformation information, further crystallographic analysis was conducted in this work. Figure 8a,b show the phase maps (zinc-rich crystals—red; β-Sn matrix—blue) and the all-Euler orientation micrographs corresponding to the longitudinal structures without and with a 12 T HMF, respectively. It can be seen that the large primary zinc-rich crystals are distributed in the fine β-Sn grains. Irrespective of whether the HMF is applied, the primary zinc-rich crystals show different colors, indicating that they have different orientations. However, the crystallographic calculation suggests that the HMF tends to orient its c-axes perpendicular to **B**. Figure 9a,b show the scattered <0001> pole figures corresponding to the primary zinc-rich crystals in Figure 8a,b, respectively. The <0001> poles are randomly distributed in the absence of the HMF, whereas they reveal a linear distribution in the presence of the HMF. This further proves that the primary zinc-rich crystals tend to orient preferentially with the <0001> direction, i.e., the c-axis, perpendicular to **B**. This orientation feature is consistent with that of the dendritic primary zinc-rich crystals under a HMF [13]. A detailed crystallographic analysis was also conducted on the β-Sn matrix, implying no preferential orientation, even with a 12 T HMF.

As analysed in other work [13], the preferential orientation of the hexagonal primary zinc-rich crystals should be related to their magnetocrystalline anisotropy. According to the magnetization energy theory, the easy magnetization axis should be parallel to the magnetic field for paramagnetic materials and perpendicular to the magnetic field for diamagnetic materials [18,28]. Zinc is a diamagnetic material with a smaller magnetic susceptibility along the c-axis (-0.169×10^{-6} cm^3/g) than that in the direction perpendicular to the c-axis (0.124×10^{-6} cm^3/g) [29]. To minimize the magnetization energy, the c-axis of the primary zinc-rich crystals tends to be rotated to the direction perpendicular to **B** under the HMFs. The driving force to start the rotation is the magnetic torque, as analysed in detail elsewhere [17,18,30]. However, to complete this rotation, a weak constraint medium is indispensable for the crystals [28]. For the present Sn-Zn-Bi alloy in this work, the zinc-rich crystals are primarily crystallized from the melt during the solidification process. The surrounding liquid provides a free rotation environment for them. However, for the crystals in the peripheral regions of the specimens, their rotations may be hindered by the crucible walls. It is known that the melt solidifies inward from the crucible wall in a non-directional solidification way. This means that the crucible walls may act as the effective nucleation sites for the primary zinc-rich crystals, as a result of which their rotations are prevented by the crucible walls. After nucleation, these crystals grow towards the central regions of the specimens, and finally reveal random alignments in all cases.

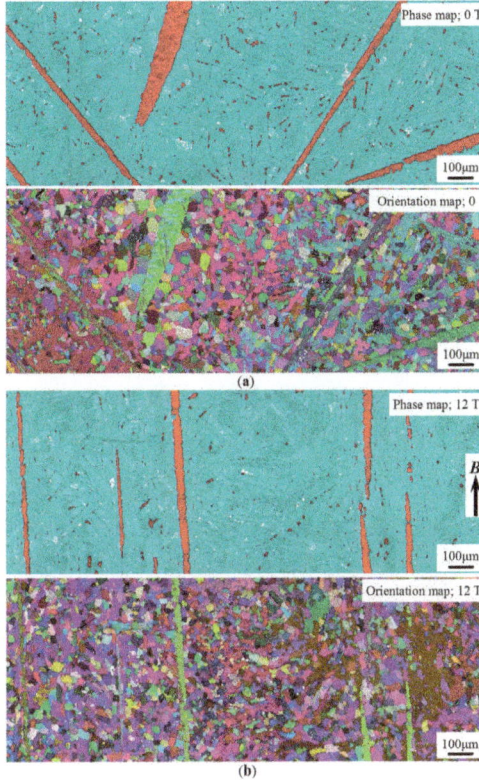

Figure 8. Phase maps and all-Euler orientation micrographs corresponding to the longitudinal structures (**a**) without and (**b**) with a 12 T HMF, respectively. The arrow indicates the direction of **B**.

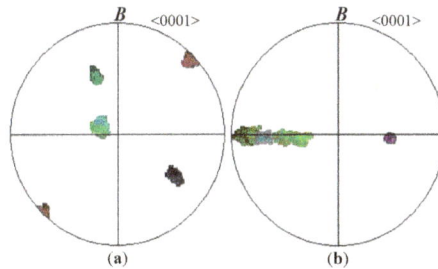

Figure 9. Scattered <0001> pole figures corresponding to the primary zinc-rich crystals in (**a**) Figure 8a and (**b**) in Figure 8b, respectively.

Now it can be understood that in the absence of the HMFs, all the plate-shaped primary zinc-rich crystals align randomly, as schematically shown in Figure 10a. When the HMFs are applied, the *c*-axes of the crystals in the central regions of specimens are fixed in the plane perpendicular to *B*. As the {0001} basal plane corresponds to the large surfaces of the plate-shaped crystals, they finally exhibit a vertical alignment, as schematically shown in Figure 10b.

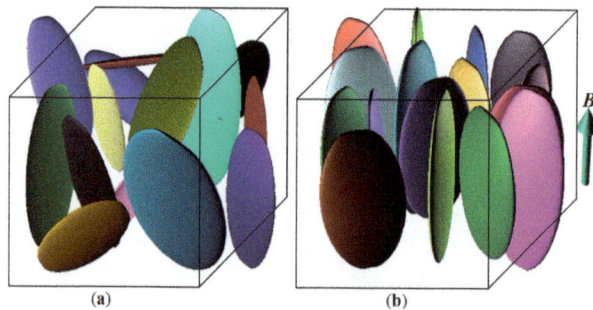

(a)   (b)

Figure 10. Schematic diagrams showing the alignments of the plate-shaped primary zinc-rich crystals (a) without and (b) with a HMF, respectively. The arrow in (b) indicates the direction of *B*.

3. Materials and Methods

A ternary Sn-12 wt % Zn-6 wt % Bi (nominal composition) alloy was prepared by melting pure zinc (99.9 wt %), tin (99.95 wt %), and bismuth (99.9 wt %) in an induction furnace. The alloy was reheated to 673 K (400 °C) at a rate of 5 K/min in an electric resistance furnace under an argon atmosphere and held at this temperature for 25 min to ensure a homogeneous composition. Afterwards, the melt was cooled to room temperature at a rate of about 10 K/min. From the start of the reheating until the end of the cooling, different uniform-distributed HMFs were applied, as introduced in detail elsewhere [13].

The HMF-treated as-cast ingots (Φ 10 × 12 mm) were cut longitudinally or transversely (parallel or perpendicular to the HMF direction *B*), following a standard mechanical polishing. The macro/microstructures were observed with a Leica DMR optical microscope (Leica Microsystems, Wetzlar, Germany). Moreover, two small cubes were also cut from the 0 and 12 T specimens for three-dimensional microstructure observations. After further polishing plus Ar ion beam cleaning, EBSD auto scanning was performed on the specimens for crystallographic analysis on a Zeiss ULTRA PLUS FE-SEM (Carl Zeiss, Jena, Germany), equipped with an Oxford-HKL Channel 5 system (Oxford Instruments, Abingdon, Oxfordshire, UK). Differential scanning calorimetry (DSC) was carried out on a Q100 TA Instrument (TA Instruments, Newcastle, DE, USA) to understand the cooling process of the alloy: a 20 mg specimen was cooled from 623 K (350 °C) to 293 K (20 °C) at a constant rate of 5 K/min in Ar flow.

The large faceted interface of the primary zinc-rich crystals exposed to the Sn matrix was calculated by an indirect two-trace method [31,32]. The preferential direction of the primary zinc-rich crystals relative to the HMF direction is determined as follows: first, the *B* vector in the sample coordinate system is transformed to its equivalent vector in the crystal coordinate system with the measured orientation data of the crystals; then, a plane is defined by the cross product of these two differently oriented vectors; finally, the statistically representative crystalline direction close to the normal vector of the plane is determined as the preferential direction [13].

4. Conclusions

A ternary Sn-12Zn-6Bi alloy was solidified under different HMFs, and it was found that:

a. Irrespective of whether the HMF is applied, the primary zinc-rich crystals reveal a plate-shaped form and grow in three dimensions, as follows: spreading rapidly in the {0001} basal plane and then decreasing the thickness at the edges gradually;

b. In the absence of the HMF, the plate-shaped primary zinc-rich crystals align randomly, whereas in the presence of the HMF, those in the central regions of the specimens tend to align vertically;

c. Crystallographic analyses indicate that the vertically aligned primary zinc-rich crystals orient preferentially with the *c*-axis perpendicular to *B*;

d. The preferential orientation and thus the regular vertical alignment should be related to the magnetocrystalline anisotropy of the diamagnetic primary zinc-rich crystals.

Acknowledgments: The authors acknowledge the Liaoning Provincial Natural Science Foundation of China (2015021002), the Fundamental Research Funds for the Central Universities (N150904003), the China Postdoctoral Science Foundation (2015M570250), the Northeastern University Postdoctoral Science Foundation (20150202), the National Natural Science Foundation of China (51690161, 51201029, 51574075, 51374067 and 51674078), and the Science and Technology Program of Guangzhou, China (2015B090926013).

Author Contributions: Chunyan Ban conceived and designed the experiments; Lei Li wrote the paper; Ruixue Zhang performed the experiments; Haitao Zhang, Minghui Cai and Yubo Zuo contributed the materials; Qingfeng Zhu, Xiangjie Wang, and Jianzhong Cui conducted the data analysis.

Conflicts of Interest: The authors declare no conflict of interest.

References

1. Li, L.; Zhang, Y.D.; Esling, C.; Zhao, Z.H.; Zuo, Y.B.; Zhang, H.T.; Cui, J.Z. Formation of feathery grains with the application of a static magnetic field during direct chill casting of Al-9.8 wt % Zn alloy. *J. Mater. Sci.* **2009**, *44*, 1063–1068. [CrossRef]

2. Li, L.; Zhang, Y.D.; Esling, C.; Zhao, Z.H.; Zuo, Y.B.; Zhang, H.T.; Cui, J.Z. Formation of twinned lamellas with the application of static magnetic fields during semi-continuous casting of Al-0.24 wt % Fe alloy. *J. Cryst. Growth* **2009**, *311*, 3211–3215. [CrossRef]

3. Vives, C. Effects of forced electromagnetic vibrations during the solidification of aluminum alloys: Part II. solidification in the presence of colinear variable and stationary magnetic fields. *Metall. Trans. B* **1996**, *27*, 457–464. [CrossRef]

4. Vives, C. Effects of forced electromagnetic vibrations during the solidification of aluminum alloys: Part I. solidification in the presence of crossed alternating electric fields and stationary magnetic fields. *Metall. Trans. B* **1996**, *27*, 445–455. [CrossRef]

5. Singh, R.; Thomas, B.G.; Vanka, S.P. Effects of a magnetic field on turbulent flow in the mold region of a steel caster. *Metall. Trans. B* **2013**, *44*, 1201–1221. [CrossRef]

6. Hu, S.D.; Dai, Y.C.; Gagnoud, A.; Fautrelle, Y.; Moreau, R.; Ren, Z.M.; Deng, K.; Li, C.J.; Li, X. Effect of a magnetic field on macro segregation of the primary silicon phase in hypereutectic Al-Si alloy during directional solidification. *J. Alloy. Compd.* **2017**, *722*, 108–115. [CrossRef]

7. Li, H.X.; Fautrelle, Y.; Hou, L.; Du, D.F.; Zhang, Y.K.; Ren, Z.M.; Lu, X.G.; Moreau, R.; Li, X. Effect of a weak transverse magnetic field on the morphology and orientation of directionally solidified Al–Ni alloys. *J. Cryst. Growth* **2016**, *436*, 68–75. [CrossRef]

8. Du, D.F.; Guan, G.; Gagnoud, A.; Fautrelle, Y.; Ren, Z.M.; Lu, X.G.; Wang, H.; Dai, Y.M.; Wang, Q.L.; Li, X. Effect of a high magnetic field on the growth of å-CuZn5 dendrite during directionally solidified Zn-rich Zn–Cu alloys. *Mater. Charact.* **2016**, *111*, 31–42. [CrossRef]

9. Li, X.; Wang, J.; Du, D.F.; Zhang, Y.K.; Fautrelle, Y.; Nguyen-Thi, H.; Gagnoud, A.; Ren, Z.M.; Moreau, R. Effect of a transverse magnetic field on the growth of equiaxed grains during directional solidification. *Mater. Lett.* **2015**, *161*, 595–600. [CrossRef]

10. Li, L.; Zhu, Q.F.; Zhang, H.; Zuo, Y.B.; Ban, C.Y.; He, L.Z.; Liu, H.T.; Cui, J.Z. Morphological and crystallographic characterization of solidified Al–3Ti–1B master alloy under a high magnetic field. *Mater. Charact.* **2014**, *95*, 1–11. [CrossRef]

11. Xuan, W.D.; Liu, H.; Lan, J.; Li, C.J.; Zhong, Y.B.; Li, X.; Cao, G.H.; Ren, Z.M. Effect of a transverse magnetic field on stray grain formation of Ni-Based single crystal superalloy during directional solidification. *Metall. Trans. B* **2016**, *6*, 3231–3236. [CrossRef]

12. Xuan, W.D.; Liu, H.; Li, C.J.; Ren, Z.M.; Zhong, Y.B.; Li, X.; Cao, G.H. Effect of a high magnetic field on microstructures of Ni-Based single crystal superalloy during seed melt-back. *Metall. Trans. B* **2016**, *47*, 828–833. [CrossRef]

13. Li, L.; Li, Z.B.; Zhang, Y.D.; Esling, C.; Liu, H.T.; Zhao, Z.H.; Zhu, Q.F.; Zuo, Y.B.; Cui, J.Z. Crystallographic effect of a high magnetic field on hypoeutectic Zn-Al during solidification process. *J. Appl. Crystallogr.* **2014**, *47*, 606–612. [CrossRef]

14. Li, L.; Ban, C.Y.; Shi, X.C.; Zhang, H.T.; Cai, M.H.; Liu, H.T.; Cui, J.Z.; Nagaumi, H. Crystallographic growth pattern of zinc-rich plate-like cells under a high magnetic field. *Mater. Lett.* **2016**, *185*, 447–451. [CrossRef]

15. Li, L.; Zhang, Y.D.; Esling, C.; Jiang, H.X.; Zhao, Z.H.; Zuo, Y.B.; Cui, J.Z. Influence of a high magnetic field on the precipitation behaviors of the primary Al_3Fe phase during the solidification of hypereutectic Al-3.31 wt % Fe alloy. *J. Cryst. Growth* **2012**, *339*, 61–69. [CrossRef]

16. Li, L.; Zhao, Z.H.; Zuo, Y.B.; Zhu, Q.F.; Cui, J.Z. Effect of a high magnetic field on the morphological and crystallographic features of primary Al_6Mn phase formed during solidification process. *J. Mater. Res.* **2013**, *28*, 1567–1573. [CrossRef]

17. Li, L.; Zhang, Y.D.; Esling, C.; Qin, K.; Zhao, Z.H.; Zuo, Y.B.; Cui, J.Z. A microstructural and crystallographic investigation on the precipitation behaviour of primary Al_3Zr phase under high magnetic field. *J. Appl. Crystallogr.* **2013**, *46*, 421–429. [CrossRef]

18. Wang, C.J.; Wang, Q.; Wang, Z.Y.; Li, H.T.; Nakajima, K.; He, J.C. Phase alignment and crystal orientation of Al_3Ni in Al–Ni alloy by imposition of a uniform high magnetic field. *J. Cryst. Growth* **2008**, *310*, 1256–1263. [CrossRef]

19. Li, X.; Ren, Z.M.; Fautrelle, Y. The alignment, aggregation and magnetization behaviors in MnBi–Bi composites solidified under a high magnetic field. *Intermetallics* **2007**, *15*, 845–855. [CrossRef]

20. Li, X.; Ren, Z.M.; Fautrelle, Y. Effect of high magnetic fields on the microstructure in directionally solidified Bi–Mn eutectic alloy. *J. Cryst. Growth* **2007**, *299*, 41–47. [CrossRef]

21. Li, L.; Xu, B.; Tong, W.P.; He, L.Z.; Ban, C.Y.; Zhang, H.; Zuo, Y.B.; Zhu, Q.F.; Cui, J.Z. Directional growth behavior of α(Al) dendrites during concentration-gradient-controlled solidification process in static magnetic field. *Trans. Nonferrous Met. Soc. China* **2015**, *25*, 2438–2445. [CrossRef]

22. Du, D.F.; Fautrelle, Y.; Ren, Z.M.; Moreau, R.; Li, X. Effect of a high magnetic field on the growth of ternary Al–Cu–Ag alloys during directional solidification. *Acta Mater.* **2016**, *121*, 240–256. [CrossRef]

23. Li, X.; Fautrelle, Y.; Ren, Z.M. Influence of thermoelectric effects on the solid–liquid interface shape and cellular morphology in the mushy zone during the directional solidification of Al–Cu alloys under a magnetic field. *Acta Mater.* **2007**, *55*, 3803–3813. [CrossRef]

24. Rhême, M.; Gonzales, F.; Rappaz, M. Growth directions in directionally solidified Al–Zn and Zn–Al alloys near eutectic composition. *Scripta Mater.* **2008**, *59*, 440–443.

25. Haxhimali, T.; Karma, A.; Gonzales, F.; Rappaz, M. Orientation selection in dendritic evolution. *Nat. Mater.* **2006**, *5*, 660–664. [CrossRef] [PubMed]

26. Li, L.; Ban, C.Y.; Shi, X.C.; Zhang, H.T.; Zuo, Y.B.; Zhu, Q.F.; Wang, X.J.; Zhang, H.; Cui, J.Z.; Nagaumi, H. Effects of a high magnetic field on the primary zinc-rich crystals in hypoeutectic Zn–Sn alloy. *J. Cryst. Growth* **2017**, *463*, 59–66. [CrossRef]

27. Kima, Y.S.; Kim, K.S.; Hwang, C.W.; Suganuma, K. Effect of composition and cooling rate on microstructure and tensile properties of Sn–Zn–Bi alloys. *J. Alloy. Compd.* **2003**, *352*, 237–245. [CrossRef]

28. Asai, S. Crystal orientation of non-magnetic materials by imposition of a high magnetic field. *Sci. Technol. Adv. Mater.* **2003**, *4*, 455–460. [CrossRef]

29. Marcus, J.A. Magnetic Susceptibility of Zinc at Liquid Hydrogen Temperatures. *Phys. Rev.* **1949**, *76*, 413–416. [CrossRef]

30. Sugiyama, T.; Tahashi, M.; Sassa, K.; Asai, S. The control of crystal orientation in non-magnetic metals by imposition of a high magnetic field. *ISIJ. Int.* **2003**, *43*, 855–861. [CrossRef]

Crystals **2017**, *7*, 204

31. Zhang, Y.D.; Esling, C.; Zhao, X.; Zuo, L. Indirect two-trace method to determine a faceted low-energy interface between two crystallographically correlated crystals. *J. Appl. Cryst.* **2007**, *40*, 436–440. [CrossRef]
32. Li, L.; Zhang, Y.D.; Esling, C.; Jiang, H.; Zhao, Z.H.; Zuo, Y.B.; Cui, J.Z. Crystallographic features of primary Al$_3$Zr phase. *J. Cryst. Growth* **2011**, *316*, 172–176. [CrossRef]

crystals

MDPI

Article

The Mixed-Metal Oxochromates(VI) Cd(Hg$^I_2$)$_2$(Hg^{II})$_3$O$_4$(CrO$_4$)$_2$, Cd(Hg^{II})$_4$O$_4$(CrO$_4$) and Zn(Hg^{II})$_4$O$_4$(CrO$_4$)—Examples of the Different Crystal Chemistry within the Zinc Triad

Matthias Weil

Institute for Chemical Technologies and Analytics, Division of Structural Chemistry, TU Wien, Getreidemarkt 9/164-SC, A-1060 Vienna, Austria; Matthias.Weil@tuwien.ac.at; Tel.: +43-1-58801-17122

Academic Editor: Helmut Cölfen
Received: 11 October 2017; Accepted: 3 November 2017; Published: 6 November 2017

Abstract: The three mixed-metal oxochromates(VI) Cd(Hg$^I_2$)$_2$(Hg^{II})$_3$O$_4$(CrO$_4$)$_2$, Cd(Hg^{II})$_4$O$_4$(CrO$_4$), and Zn(Hg^{II})$_4$O$_4$(CrO$_4$) were grown under hydrothermal conditions. Their crystal structures were determined from single-crystal X-ray diffraction data. The crystal-chemical features of the respective metal cations are characterised, with a linear coordination for mercury atoms in oxidation states +I and +II, octahedral coordination spheres for the divalent zinc and cadmium cations and a tetrahedral configuration of the oxochromate(VI) anions. In the crystal structures the formation of two subunits is apparent, viz. a mercury-oxygen network and a network of cadmium (zinc) cations that are directly bound to the oxochromate(VI) anions. An alternative description of the crystal structures based on oxygen-centred polyhedra is also given.

Keywords: zinc; cadmium; mercury; oxochromates(VI); crystal chemistry; oxo-centred polyhedra

1. Introduction

The three elements of the zinc triad have a closed-shell $nd^{10}(n + 1)s^2$ electronic configuration with n = 3, 4, and 5 for zinc, cadmium, and mercury, respectively. In compounds of these elements with ionic or predominantly ionic character, zinc exclusively exhibits oxidation state +II, cadmium with very few exceptions has an oxidation state of +II (Cd$_2$(AlCl$_4$)$_2$ being one of them with an oxidation state of +I [1,2]), whereas a multitude of mercuric (oxidation state +II), mercurous (oxidation state +I) and mixed-valent mercury compounds are known. The crystal-chemical features of all three elements are remarkably different. The most frequently observed coordination numbers for zinc in its compounds are 4, 5, and 6 with (distorted) tetrahedral, trigonal-bipyramidal, and octahedral coordination environments, respectively. The larger cadmium cation has a coordination number of four only in combination with larger anions (like in CdS), and in the majority of cases exhibits coordination numbers of six, or higher. For most of the latter cases, the coordination spheres are considerably distorted and difficult to derive from simple polyhedra. In many aspects, including structural characteristics, zinc and cadmium compounds resemble their alkaline earth congeners magnesium and calcium, respectively, which likewise have a closed shell electronic configuration. Mercury, on the other hand, is unique amongst all metals (*cf.* the low melting point) and has a peculiar crystal chemistry, showing a preference for linear coordination by more electronegative elements (coordination number of two). To a certain extent, these features can be related to relativistic effects that are pronounced for this element [3,4]. While a number of review articles devoted to the crystal chemistry of mercury have been published over the years [5–11], to the best of the author's knowledge, apart from chapters in a compendium on coordination chemistry [11,12], special reviews on the crystal chemisty of zinc or cadmium did not appear thus far.

During previous crystal growth experiments it was successfully shown that mixed-metal compounds of the zinc triad can be prepared under hydrothermal conditions in form of their sulfate or selenate salts, viz. $CdXO_4(HgO)_2$ (X = S, Se) [13], $(MXO_4)_2(HgO)_2(H_2O)$ (X = S, Se; M = Cd, Zn), $CdSeO_4(Hg(OH)_2)$, and $(ZnSe^{IV}O_3)(ZnSe^{VI}O_4)Hg^I_2(OH)_2$ [14]. In the present study it was intended to replace the sulfate (SO_4^{2-}) or selenate (SeO_4^{2-}) anions with chromate anions (CrO_4^{2-}) to search for new mixed-metal compounds of the zinc triad. Chromates, in particular, appeared to be promising candidates for formation of new compounds because they show pH-dependent chromate $\rightleftharpoons$ dichromate equilibria and are able to stabilize different oxidation states for mercury. Mercurous chromates(VI) are scarce and known only for dimorphic Hg_2CrO_4 [15] and $Hg_6Cr_2O_9$ [16], whereas mercuric chromates are more frequent with structure determinations reported for dimorphic $HgCrO_4$ [17,18], for $Hg_3O_2CrO_4$ [19], $HgCr_2O_7$ [20], $HgCrO_4(H_2O)_{0.5}$ [21], and $HgCrO_4(H_2O)$ [18]. In addition to these mercurous and mercuric chromates(VI), the mixed-valent Hg(I/II) compounds $(Hg^I_2)_2O(CrO_4)(Hg^{II}O)$ (mineral name wattersite [22]) and $Hg_6Cr_2O_{10}$ (=$2Hg_2CrO_4 \cdot 2HgO$) [16] are also known. The two lead(II) mercury(II) chromates(VI) Pb_2HgCrO_6 [23] and $Pb_2(Hg_3O_4)(CrO_4)$ [24] served as a proof of concept that additional metal ions can be incorporated into mercury oxochromates(VI). Crystallographic data for zinc and cadmium chromates, on the other hand, are restricted to $CrVO_4$-type $ZnCrO_4$ [25], $Zn_2(OH)_2CrO_4$ [26], and to dimorphic $CdCrO_4$ (low-temperature form, *Cmcm*; high-temperature form, *C2/m*) and Cd_2CrO_5 [27], respectively.

2. Results and Discussion

Three mixed-metal oxochromates(VI) were obtained under the given hydrothermal conditions, viz. $Cd(Hg^I_2)_2(Hg^{II})_3O_4(CrO_4)_2$, $Cd(Hg^{II})_4O_4(CrO_4)$ and $Zn(Hg^{II})_4O_4(CrO_4)$. Although the educt ratio Hg:Cd(Zn):Cr was 2:1:1, the ratio in the solid reaction products was different with a much higher mercury content, namely 7:1:2 for $Cd(Hg^I_2)_2(Hg^{II})_3O_4(CrO_4)_2$, 4:1:1 for $Cd(Hg^{II})_4O_4(CrO_4)$ and $Zn(Hg^{II})_4O_4(CrO_4)$, and 5:0:1 for wattersite crystals. The formation of mixed-valent mercury(I,II) compounds, i.e., wattersite in both batches and $Cd(Hg^I_2)_2(Hg^{II})_3O_4(CrO_4)_2$ in the cadmium-containing batch, indicates that complex redox equilibria between different mercury species (Hg(0) $\rightleftharpoons$ Hg(I) $\rightleftharpoons$ Hg(II)) must have been present under the chosen hydrothermal reaction conditions. Such redox equilibria are easily influenced by the presence of additional redox partners, here, for example Cr(VI) $\rightleftharpoons$ Cr(III), and other interacting parameters like temperature, pressure, pH, concentration of the reactants, etc. Such a complex interplay between different adjustable parameters not only makes a prediction of solid products difficult, but can also lead to multi-phase formation and the presence of element species with different oxidation states in one batch. This kind of behaviour is not only exemplified by the three title compounds but also for other mixed-valent mercury oxocompounds that were obtained under similar hydrothermal conditions [16,28–30].

The strong preference for linear coordination of mercuric and mercurous cations is confirmed in the crystal structures of the three title compounds where O–Hg–O and/or Hg–Hg–O units with Hg–O bond lengths less than 2.2 Å are present. Representative bond lengths of the three title compounds are listed in Table 1.

Table 1. Selected bond lengths (Å) and angles (°).

Cd(Hg$^I_2$)$_2$(Hg^{II})$_3$O$_4$(CrO$_4$)$_2$			Zn(Hg^{II})$_4$O$_4$(CrO$_4$)		
Hg1	O4	2.002(8)	Hg1	O3	2.030(7)
Hg1	O5	2.016(8)	Hg1	O4	2.045(6)
Hg1	O3	2.732(11)	Hg1	O8	2.703(7)
Hg1	O2	2.734(13)	Hg1	O1	2.805(8)
Hg2	O6	2.192(8)	Hg1	O4	2.819(7)
Hg2	O5	2.528(8)	Hg1	O7	2.840(7)
Hg2	Hg3	2.5301(6)	Hg2	O1	2.043(6)
Hg2	O4	2.692(9)	Hg2	O2	2.069(6)
Hg3	O4	2.098(8)	Hg2	O4	2.728(7)

Table 1. *Cont.*

Cd(Hg$^I_2$)$_2$(Hg^{II})$_3$O$_4$(CrO$_4$)$_2$				Zn(Hg^{II})$_4$O$_4$(CrO$_4$)			
Hg3	O1	2.734(10)		Hg2	O5	2.776(7)	
Hg3	O1	2.803(11)		Hg2	O5	2.896(7)	
Hg4	O5	2.037(8)	2x	Hg2	O8	2.903(7)	
Hg4	O6	2.600(9)	2x	Hg3	O3	2.015(6)	
Cd	O5	2.252(7)	2x	Hg3	O4	2.024(6)	
Cd	O4	2.293(9)	2x	Hg3	O7	2.610(7)	
Cd	O2	2.322(11)	2x	Hg3	O8	2.838(8)	
Cr	O1	1.611(11)		Hg3	O4	2.932(7)	
Cr	O3	1.615(10)		Hg4	O1	2.009(6)	
Cr	O2	1.665(12)		Hg4	O2	2.027(6)	
Cr	O6	1.697(8)		Hg4	O6	2.625(7)	
				Hg4	O8	2.731(8)	
O4	Hg1	O5	175.2(3)	Hg4	O7	2.746(7)	
O5	Hg4	O5	180.0	Zn	O3	2.045(8)	
O6	Hg2	Hg3	165.6(2)	Zn	O1	2.055(6)	
Hg3	Hg2	O4	94.91(17)	Zn	O2	2.075(6)	
				Zn	O5	2.097(6)	
	Cd(Hg^{II})$_4$O$_4$(CrO$_4$)			Zn	O6	2.146(6)	
Hg1	O4	2.016(7)		Zn	O2	2.325(7)	
Hg1	O3	2.049(6)		Cr	O8	1.634(7)	
Hg1	O7	2.638(7)		Cr	O7	1.643(7)	
Hg1	O2	2.667(6)		Cr	O6	1.652(7)	
Hg1	O7	2.790(7)		Cr	O5	1.657(6)	
Hg2	O2	2.012(6)					
Hg2	O1	2.045(7)		O3	Hg1	O4	172.8(3)
Hg2	O5	2.584(7)		O1	Hg2	O2	163.3(3)
Hg2	O7	2.740(7)		O3	Hg3	O4	176.6(3)
Hg2	O8	2.882(8)		O1	Hg4	O2	175.7(3)
Hg3	O1	2.057(6)		Hg4	O1	Hg2	115.1(3)
Hg3	O2	2.062(6)		Hg4	O2	Hg2	116.0(3)
Hg3	O4	2.577(6)		Hg3	O3	Hg1	123.2(4)
Hg3	O8	2.725(7)		Hg3	O4	Hg1	120.2(3)
Hg3	O6	2.752(7)					
Hg3	O4	2.838(8)					
Hg4	O4	2.014(7)					
Hg4	O3	2.026(6)					
Hg4	O8	2.700(7)					
Hg4	O4	2.838(8)					
Cd	O3	2.237(6)					
Cd	O5	2.251(7)					
Cd	O2	2.273(6)					
Cd	O6	2.283(7)					
Cd	O1	2.299(6)					
Cd	O1	2.421(6)					
Cr	O8	1.620(7)					
Cr	O7	1.627(7)					
Cr	O6	1.633(7)					
Cr	O5	1.658(7)					
O4	Hg1	O3	173.6(3)				
O2	Hg2	O1	174.0(3)				
O1	Hg3	O2	166.4(2)				
O4	Hg4	O3	176.6(3)				
Hg2	O1	Hg3	118.6(3)				
Hg2	O2	Hg3	117.0(3)				
Hg4	O3	Hg1	109.3(3)				
Hg4	O4	Hg1	122.2(3)				

The mixed-valent Cd(Hg$^I_2$)$_2$(Hg^{II})$_3$O$_4$(CrO$_4$)$_2$ phase crystallizes with one formula unit in space group $P\bar{1}$. It comprises four unique mercury cations, two of which (Hg2, Hg3) belong to a Hg$_2^{2+}$ dumbbell, and two of which (Hg1, Hg4) to Hg^{2+} cations. Hg1 is bound to two O atoms (O4, O5) at a distance of 2.002(8) and 2.016(8) Å with a nearly linear O4–Hg1–O5 angle of 175.2(3)°. Hg4, located on an inversion centre, shows two short distances of 2.037(8) Å to O5, and due to the symmetry restriction a linear O5–Hg4–O5($-x + 1$, $-y + 1$, $-z$) angle. The Hg_2^{2+} dumbbell exhibits a Hg2–Hg3 distance of 2.5301(6) Å, which is slightly above the arithmetic mean of 2.518(25) Å calculated for more than one hundred different Hg_2^{2+} dumbbells that are present in crystal structures of various inorganic oxocompounds [30]. The two O atoms tightly bonded to the Hg2–Hg3 dumbbell have distances of Hg2–O6 = 2.192(8) Å and Hg3–O4 = 2.098(8) Å but only one of them has an arrangement approaching linearity with respect to the dumbbell (O6–Hg2–Hg3 = 165.6(2)°) while the other is virtually vertical to the dumbbell (O4–Hg2–Hg3 = 94.91(17)°). Under consideration of one longer Hg3–O5 bond of 2.528(8) Å, the mercuric and mercurous cations and the three oxygen sites O4–O6 are fused into strings with the composition $\{(Hg^I_2)_2(Hg^{II})_3O_6\}^{2-}$ that are aligned into sheets extending parallel to $(01\bar{1})$ (Figure 1).

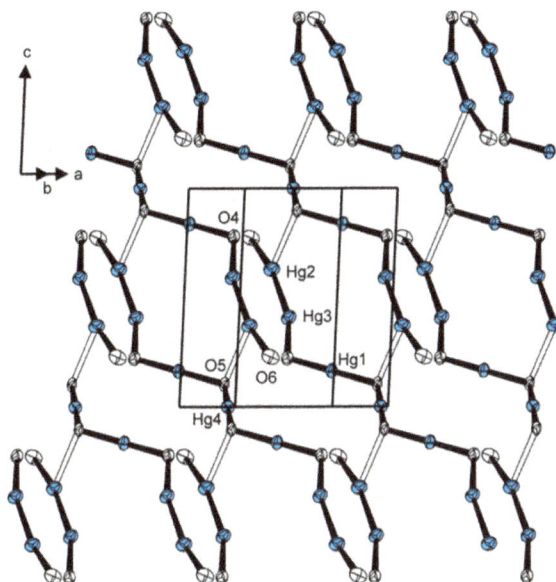

Figure 1. The Hg–O network in the structure of $Cd(Hg^I_2)_2(Hg^{II})_3O_4(CrO_4)_2$ in a projection along $[1\bar{3}0]$. Displacement ellipsoids are drawn at the 74% probability level. Short Hg–O bonds < 2.2 Å are given as solid lines, and longer Hg–O bonds as open lines.

The Cd^{2+} cation (located on an inversion centre) and the Cr(VI) atom are situated between the sheets. They are bound to six and four oxygen sites in form of slightly distorted polyhedra with octahedral and tetrahedral configurations, respectively. The $[CdO_6]$ octahedron is flanked by two $[CrO_4]$ tetrahedra sharing two corner O atoms (O2 and its symmetry-related counterpart). The range of Cd–O bond lengths in the $[CdO_6]$ octahedron is narrow (2.252(7)–2.322(11) Å), with a mean of 2.29 Å; the corresponding values for the $[CrO_4]$ tetrahedron are 1.611(11)–1.677(8) and 1.65 Å, in good agreement with typical values for oxochromates(VI) comprising isolated $[CrO_4]^{2-}$ anions (1.646(25) Å) [31]. By sharing some of the oxygen sites of the resulting $\{CdO_4(CrO_4)_2\}$ groups with the $\{(Hg^I_2)_2(Hg^{II})_3O_6\}$ network and also by additional Hg–O interactions > 2.2 Å, the three-dimensional framework structure of $Cd(Hg^I_2)_2(Hg^{II})_3O_4(CrO_4)_2$ is established (Figure 2).

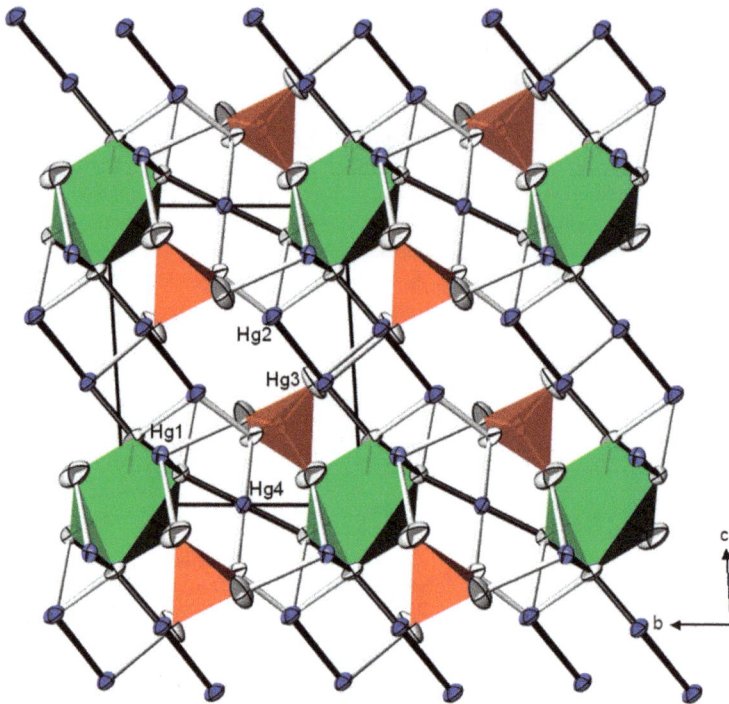

Figure 2. Crystal structure of $Cd(Hg^{I}_2)_2(Hg^{II})_3O_4(CrO_4)_2$ emphasizing the layered arrangement of the Hg–O network and the $[CdO_6]$ (green) and CrO_4 (red) polyhedra. Displacement ellipsoids are as in Figure 1.

The second cadmium-containing phase, $Cd(Hg^{II})_4O_4(CrO_4)$, and the zinc-containing phase, $Zn(Hg^{II})_4O_4(CrO_4)$, have the same formula type but are not isotypic. The cadmium compound shows orthorhombic symmetry (space group *Pbca*, eight formula units) whereas the symmetry of the zinc compound is triclinic (space group $P\overline{1}$, two formula units). Nevertheless, the general set-up of the two structures is very similar. Both structures contain two types of Hg–O chains defined by short Hg–O distances between 2.01 and 2.05 Å and more or less linear O–Hg–O angles (164–177°). The Hg–O–Hg angles in all these chains are around 120°, thus defining a zigzag arrangement. In the $Cd(Hg^{II})_4O_4(CrO_4)$ structure one of the chains, $[Hg4–O4–Hg1–O3]^1_\infty$, runs parallel [010], the other, $[Hg3–O2–Hg2–O1]^1_\infty$, runs parallel [100] (Figure 3a). In the $Zn(Hg^{II})_4O_4(CrO_4)$ structure the directions of propagation of the Hg–O chains are [100] for $[Hg2–O1–Hg4–O2]^1_\infty$ and [110] for $[Hg3–O4–Hg1–O3]^1_\infty$ (Figure 3b).

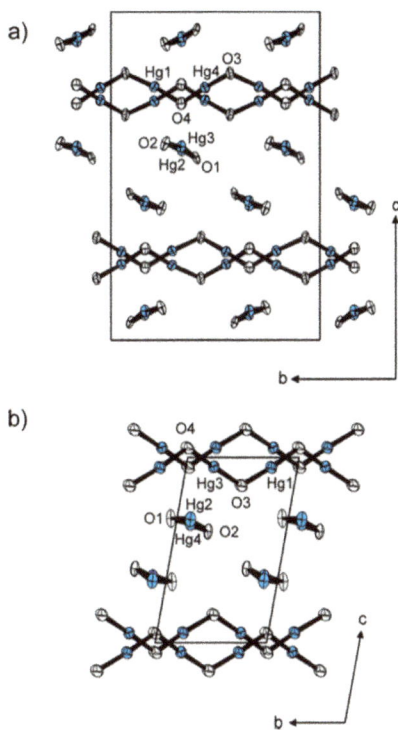

Figure 3. The two different Hg–O chains in the structures of (**a**) $Cd(Hg^{II})_4O_4(CrO_4)$ and (**b**) $Zn(Hg^{II})_4O_4(CrO_4)$. Displacement ellipsoids are drawn at the 90% probability level.

The Cd^{2+} and Zn^{2+} cations, respectively, are located between the Hg–O chains and have the function as bridging groups between adjacent Hg–O chains. Under consideration of other oxygen atoms (O5, O6) that are not part of the Hg–O chains, both metal sites have a distorted octahedral coordination environment. The Cd–O bond lengths are in a greater range than those of the $[CdO_6]$ octahedron in the structure of $Cd(Hg^I{}_2)_2(Hg^{II})_3O_4(CrO_4)_2$, 2.237(6)–2.421(6) Å, but have the same mean value of 2.29 Å. The Zn–O bond lengths in $Zn(Hg^{II})_4O_4(CrO_4)$ are expectedly shorter (2.045(8)–2.325(7) Å; mean 2.12 Å). In both $M(Hg^{II})_4O_4(CrO_4)$ structures (M = Cd, Zn) two $[MO_6]$ octahedra are fused via edge-sharing into a $[M_2O_{10}]$ double octahedron. These double octahedra are aligned in layers parallel (001) and have the same orientation in each layer in the structure of $Zn(Hg^{II})_4O_4(CrO_4)$ (Figure 4), whereas their orientations alternate in the structure of $Cd(Hg^{II})_4O_4(CrO_4)$ due to the presence of the *a* glide plane (Figure 5).

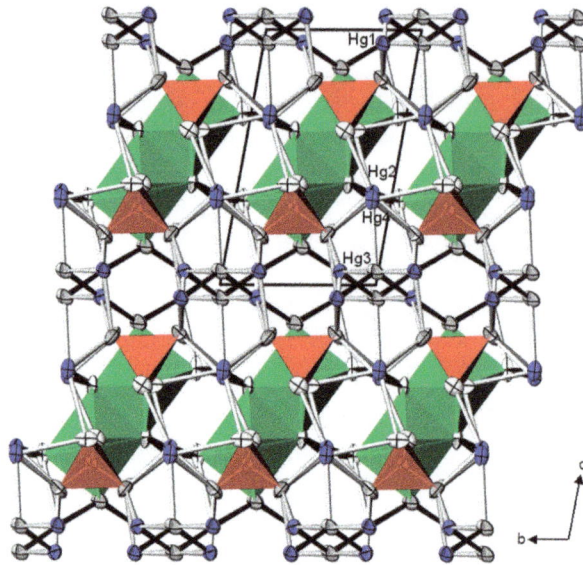

Figure 4. The crystal structure of $Zn(Hg^{II})_4O_4(CrO_4)$. $[CrO_4]$ tetrahedra are red, $[ZnO_6]$ octahedra are green. Displacement ellipsoids are as in Figure 3.

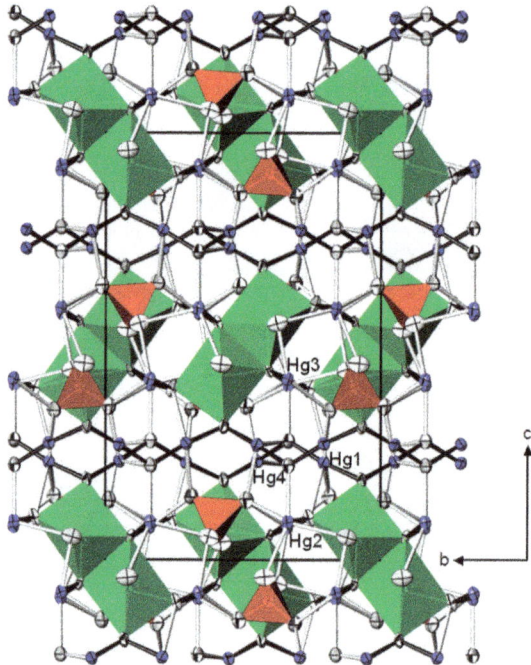

Figure 5. The crystal structure of $Cd(Hg^{II})_4O_4(CrO_4)$. $[CrO_4]$ tetrahedra are red, and $[CdO_6]$ octahedra are green. Displacement ellipsoids are as in Figure 3.

The Cr(VI) atoms sit above and below the $[M_2O_{10}]$ double octahedra and link them through two bridging vertex O atoms into "MCrO$_4$" (M = Cd (Zn)) slabs extending parallel [100]. The structural characteristics of the tetrahedral $[CrO_4]$ groups in the two structures follow the general trend [31] and in direct comparison show subtle differences. A somewhat greater distortion for the cadmium-containing structure (1.620(7)–1.658(7) Å, 108.8(4)–111.0(4)°) is observed compared to the zinc-containing structure (1.634(7)–1.657(6) Å, 108.5(4)–110.9(4)°).

The presence of two distinct structural subunits in each of the $Cd(Hg^I_2)_2(Hg^{II})_3O_4(CrO_4)_2$ and $M(Hg^{II})_4O_4(CrO_4)$ structures, viz., a mercury-oxygen network and cadmium/zinc cations bound directly to $[CrO_4]^{2-}$ anions, allows to reformulate them as $[\{(Hg^I_2)_2(Hg^{II})_3O_4\}^{2+}\{Cd(CrO_4)_2\}^{2-}]$ and $MCrO_4 \cdot 4HgO$ (M = Cd, Zn), respectively. The alternative formulae also emphasize the "basic" character (in an acid/base sense) of these compounds which is associated with the presence of oxygen atoms that are exclusively bonded to metal cations, here, those of mercury, cadmium (zinc), or mixtures thereof. Since these oxygen atoms do not belong to a chromate anion they are defined as "basic". In the vast majority of cases, such "basic" oxygen atoms are surrounded by four metal cations in the form of distorted tetrahedra. Krivovichev and co-workers have resumed the use of such oxygen-centred $[OM_4]$ tetrahedra for a rational structure description and classification of mineral and synthetic lead(II) oxo-compounds [32]. A general review of anion-centred $[OM_4]$ tetrahedra in the structures of inorganic compounds with different metals M has been published some time ago, including $[OHg_4]$ tetrahedra [33]. However, mixed $[OM_4]$ tetrahedra with M = Hg and Cd or Zn are unknown so far.

In the structure of $Cd(Hg^I_2)_2(Hg^{II})_3O_4(CrO_4)_2$, the "basic" oxygen atoms are represented by O4 and O5, both being bound to three mercury cations and one cadmium cation. The two types of $[OHg_3Cd]$ tetrahedra are considerably distorted, with O–M distances between 2.002(8) and 2.692(9) Å and M–O–M angles ranging from 98.6(3) to 123.5(4)°. Based on the alternative description by using oxygen-centred polyhedra, the $[OHg_3Cd]$ tetrahedra are linked through common edges (Cd—Hg2) and corners (Cd, Hg1, Hg4) into sheets with a width of two tetrahedra parallel (001). Adjacent sheets are connected along [001] through the Hg–Hg bond of the Hg2–Hg3 dumbbell. The remaining $[CrO_4]$ tetrahedra are situated in the voids of this arrangement and connected to the "basic" metal-oxygen network through additional Cd–O and Hg–O bonds (Figure 6).

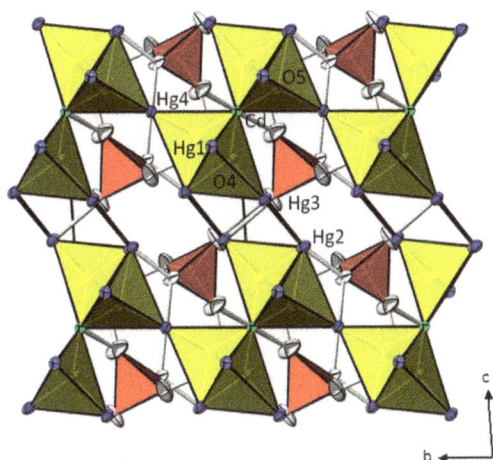

Figure 6. Crystal structure of $Cd(Hg^I_2)_2(Hg^{II})_3O_4(CrO_4)_2$ using oxygen-centred $[OHg_3Cd]$ tetrahedra (yellow) for visualisation. Displacement ellipsoids are as in Figure 1.

The "basic" O atoms in the crystal structures of $Cd(Hg^{II})_4O_4(CrO_4)$ and $Zn(Hg^{II})_4O_4(CrO_4)$ are atoms O1–O4. In the cadmium-containing structure, O1 is surrounded distorted tetrahedrally by two Hg^{2+} and two Cd^{2+} cations (bond lengths range 2.045(7)–2.421(6) Å, bond angles range 96.3(2)–118.6(3)°), O_2 from one Cd^{2+} and three Hg^{2+} cations (2.062(6)–2.667(6) Å; 88.8(2)–117.8(3)°) and O_4 from four Hg^{2+} cations (2.014(7)–2.838(8) Å; 93.1(3)–122.2(3)°). With two Hg^{2+} and one Cd^{2+} cation, O3 has only three bonding partners (2.026(6)–2.237(6) Å; 107.9(3)–119.6(3)°) that form a distorted trigonal-pyramidal polyhedron. The different types of $[OM_4]$ tetrahedra (M = Hg, Cd] and the $[OHg_2Cd]$ trigonal pyramid are linked by sharing vertices and edges into a three-dimensional framework. Like in the structure of $Cd(Hg^I_2)_2(Hg^{II})_3O_4(CrO_4)_2$, the tetrahedral $[CrO_4]$ groups in the $Cd(Hg^{II})_4O_4(CrO_4)$ structure are located in the voids of this arrangement and are connected with the framework through additional M–O bonds (Figure 7).

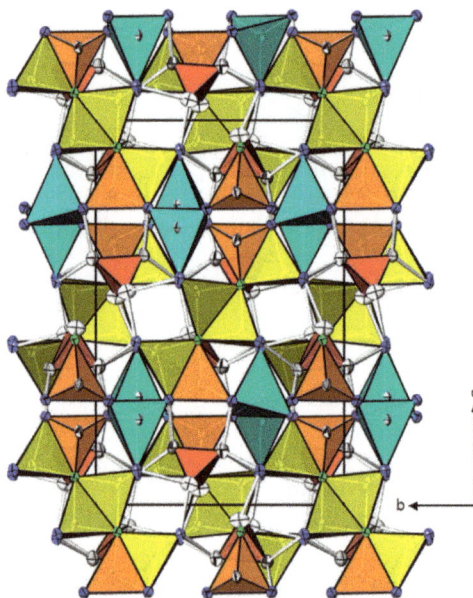

Figure 7. Crystal structure of $Cd(Hg^{II})_4O_4(CrO_4)$ using oxygen-centred tetrahedra for visualisation. $[OHg_2Cd_2]$ and $[OHg_3Cd]$ tetrahedra are yellow, $[OHg_2Cd]$ trigonal pyramids are orange and $[OHg_4]$ tetrahedra are turquoise. Displacement ellipsoids are as in Figure 3.

The above discussed similarities between the $Cd(Hg^{II})_4O_4(CrO_4)$ and $Zn(Hg^{II})_4O_4(CrO_4)$ crystal structures are also valid by using oxygen-centred polyhedra as an alternative description. The general structural set-up of $Zn(Hg^{II})_4O_4(CrO_4)$ is likewise accomplished by edge- and vertex-sharing of oxygen-centred polyhedra with $[CrO_4]$ tetrahedra in the free space and completion of the cohesion through additional M–O bonds (Figure 8). However, one of the oxygen-centered polyhedra is distinctly different. While O1 and O_2 are again surrounded tetrahedrally by Hg^{2+} and Zn^{2+} cations (2.009(6)–2.805(8) Å, 87.0(2)–123.8(3)°; 2.027(6)–2.325(7) Å, 98.8(3)–116.7(3)°), and O3 in the form of a trigonal pyramid by two Hg^{2+} and one Zn^{2+} cations (2.015(6)–2.045(8) Å, 112.5(3)–123.2(4)°), O4 has increased the number of Hg cations to which it is bound from four to five. The resulting coordination polyhedron is that of a distorted trigonal bipyramid, with the τ_5 index [34] being 0.90 [35]. The O4–Hg$_{equatorial}$ bond lengths and corresponding angles range between 2.024(6) and 2.728(7) Å and 117.8(3)–121.6(3)°, respectively; the O4–Hg$_{axial}$ bond lengths are 2.819(7) and 2.933(7) Å with an angle Hg1–O4–Hg3 of 175.5(2)°.

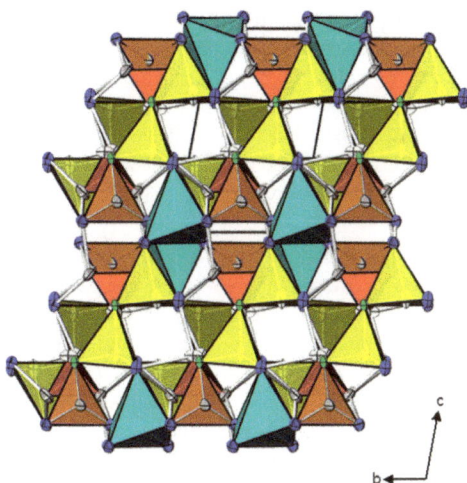

Figure 8. Crystal structure of $Zn(Hg^{II})_4O_4(CrO_4)$ using oxygen-centred tetrahedra for visualisation. $[OHg_2Cd_2]$ and $[OHg_3Cd]$ tetrahedra are yellow, $[OHg_2Cd]$ trigonal pyramids are orange and $[OHg_5]$ trigonal bipyramids are turquoise. Displacement ellipsoids are as in Figure 3.

Bond valence sums (BVS) [36], using the bond valence parameters of Brese and O'Keeffe [37], were calculated for the three structures. The results are reasonably close to the expected values (in valence sums) of 1 for mercurous Hg, 2 for mercuric Hg, 2 for Cd and Zn, 6 for Cr and 2 for O (Table 2). The global instability index GII was used as a measure of the extent to which the valence sum rule is violated [36]. The resultant GII values of 0.14 v.u. for $Cd(Hg^I_2)_2(Hg^{II})_3O_4(CrO_4)_2$, 0.14 v.u. for $Cd(Hg^{II})_4O_4(CrO_4)$ and 0.11 v.u. for $Zn(Hg^{II})_4O_4(CrO_4)$ indicate stable structures with some lattice-induced strain [38].

Table 2. Results of bond valence calculations/valence units [(1)].

$Cd(Hg^I_2)_2(Hg^{II})_3O_4(CrO_4)_2$
Hg1 2.07, Hg2 1.03, Hg3 1.04, Hg4 2.05, Cd1 2.13, Cr1 5.99, O1 1.82 [2 Hg, 1 Cr], O2 1.87 [1 Hg, 1 Cd, 1 Cr], O3 1.75 [1 Hg, 1 Cr], O4 1.94 [3 Hg, 1 Cd], O5 2.29 [3 Hg, 1 Cd], O6 1.91 [1 Cr, 2 Hg].
$Cd(Hg^{II})_4O_4(CrO_4)$
Hg1 2.13, Hg2 2.21, Hg3 2.13, Hg4 2.18, Cd1 2.12, Cr1 6.16, O1 2.21 [2 Hg, 2 Cd], O2 2.20 [3 Hg, Cd], O3 2.08 [2 Hg, Cd], O4 2.07 [4 Hg], O5 2.12 [Cr, Cd, 2 Hg], O6 2.02 [Cr, Cd, Hg], O7 1.97 [Cr, 3 Hg], O8 1.96 [Cr, 3 Hg].
$Zn(Hg^{II})_4O_4(CrO_4)$
Hg1 2.12, Hg2 2.00, Hg3 2.11, Hg4 2.19, Zn1 1.99, Cr1 5.96, O1 2.20 [3 Hg, Zn], O2 2.18 [2 Hg, 2 Zn], O3 2.15 [2 Hg, Zn], O4 1.98 [5 Hg], O5 1.99 [Cr, Zn, 2 Hg], O6 1.94 [Cr, Zn, Hg], O7 1.98 [Cr, 4 Hg], O8 1.98 [Cr, 4 Hg].

(1) For oxygen atoms the type and number of atoms they are bound to are indicated in brackets.

3. Materials and Methods

3.1. Preparation

For the hydrothermal experiments, Teflon containers with an inner volume of 5 mL were used. The metal oxides HgO, CrO_3 and ZnO (CdO), all purchased from Merck (Darmstadt, Germany), were used without further purification. 1 mmol HgO, 0.5 mmol CrO_3, and 0.5 mmol ZnO (CdO) were mixed, placed in a Teflon container and poured with 3 mL water. The container was sealed with a Teflon lid, placed in a steel autoclave, heated at 215 °C for one week and cooled within 12 h to room temperature. In both cases (cadmium- and zinc-containing batches) the final supernatant solution was colourless (pH ≈ 8), and the different crystal colours and forms indicated multi-phase formation. The solid reaction products were filtered off with a glass frit, washed with water, ethanol, and acetone and air-dried. In both the cadmium- and the zinc-containing batch, dark-red crystals of wattersite [22] were identified as the main product. In the cadmium-containing batch the two title compounds, $Cd(Hg^I_2)_2(Hg^{II})_3O_4(CrO_4)_2$ and $Cd(Hg^{II})_4O_4(CrO_4)$, were obtained as dark-red rods and orange plates, respectively, in an estimated ratio of 1:2. In the zinc-containing batch, orange plates of $Zn(Hg^{II})_4O_4(CrO_4)$ could be isolated as a minor product.

3.2. Single Crystal X-ray Diffraction

Prior to the diffraction measurements, crystals were separated from wattersite crystals and checked for optical quality under a polarizing microscope. Selected crystals were fixed with superglue on the tip of thin silica glass fibres. Intensity data were measured at room temperature with Mo-$K\alpha$ radiation, using either a SMART CCD three-circle diffractometer (Bruker, Madison, WI, USA) or a CAD-4 four-circle diffractometer with kappa geometry (Nonius, Delft, The Netherlands). After data reduction, a numerical absorption correction was performed for each data set with the aid of the HABITUS program by optimizing the crystal shape [39]. The crystal structures were solved by Direct Methods [40] and were refined using SHELXL-97 [41].

Numerical details of the data collections and structure refinements are gathered in Table 3, selected bond lengths are given in Table 1. Structure graphics were produced with ATOMS [42]. Further details of the crystal structure investigations may be obtained from the Fachinformationszentrum (Karlsruhe, Eggenstein-Leopoldshafen, Germany, Fax: +49-7247-808-666; E-Mail: crysdata@fiz-karlsruhe.de, https://www.fiz-karlsruhe.de/) on quoting the depository numbers listed at the end of Table 3.

Table 3. Details of data collections and structure refinements.

Compound		$Cd(Hg^I_2)_2(Hg^{II})_3O_4(CrO_4)_2$	$Cd(Hg^{II})_4O_4(CrO_4)$	$Zn(Hg^{II})_4O_4(CrO_4)$
Diffractometer		Siemens SMART	Nonius CAD4	Siemens SMART
Formula weight		1812.53	1094.76	1047.73
Crystal dimensions/mm^3		0.08 × 0.10 × 0.25	0.04 × 0.04 × 0.23	0.01 × 0.05 × 0.10
Crystal description		red, irregular fragment	orange, plate	yellow, plate
Space group		$P\bar{1}$	$Pbca$	$P\bar{1}$
Formula units Z		1	8	2
$a/\text{Å}$		6.1852(5)	6.9848(10)	6.873(3)
$b/\text{Å}$		7.3160(6)	12.8019(15)	6.928(3)
$c/\text{Å}$		8.5038(7)	19.227(3)	10.413(4)
$\alpha/°$		85.5840(10)	90	89.725(7)
$\beta/°$		87.2820(10)	90	70.903(7)
$\gamma/°$		72.0160(10)	90	61.694(7)
$V/\text{Å}^3$		364.80(5)	1719.3(4)	405.7(3)
μ/mm^{-1}		76.241	74.832	79.606
X-ray density/g·cm^{-3}		8.250	8.459	8.576
Range θ_{min}–θ_{max}/°		2.40–30.47	3.18–29.99	3.40–30.58
Range	h	$-8 \rightarrow 7$	$-9 \rightarrow 9$	$-9 \rightarrow 9$
	k	$-10 \rightarrow 9$	$-17 \rightarrow 17$	$-9 \rightarrow 9$
	l	$-12 \rightarrow 12$	$-27 \rightarrow 27$	$-14 \rightarrow 12$
Measured reflections		4245	18,439	4655
Independent reflections		2177	2486	2408
Obs. reflections [$I > 2\sigma(I)$]		2153	1772	1996
R_i		0.0450	0.0898	0.0431
Absorption correction			-HABITUS-	
Trans. coeff. T_{min}/T_{max}		0.004/0.055	0.1393/0.2185	0.0222/0.5407
Ext. coef. (SHELXL97)		0.0057(2)	0.000177(9)	0.00054(7)
Number of parameters		104	128	128
Δe_{max}; $\Delta e_{max}/e^-·\text{Å}^{-3}$		2.10; -1.78	1.94, -1.94	2.50; -2.24
$R[F^2 > 2\sigma(F^2)]$		0.0336	0.0244	0.0284
$wR2(F^2$ all)		0.0727	0.0446	0.0570
Goof		1.304	1.021	0.925
CSD number		433,656	433,657	433,658

4. Conclusions

During the present study it was shown that SO_4^{2-} or SeO_4^{2-} anions could be replaced with isovalent and isoconfigurational CrO_4^{2-} anions to prepare new mixed-metal oxocompounds of the zinc triad. The hydrothermally-grown crystals of $Cd(Hg^I_2)_2(Hg^{II})_3O_4(CrO_4)_2$, $Cd(Hg^{II})_4O_4(CrO_4)$, and $Zn(Hg^{II})_4O_4(CrO_4)$ each were obtained as minor reaction products in phase mixtures besides the mixed-valent mercury(I/II) compound $(Hg_2)_2O(CrO_4)(HgO)$ as the major product. All three compounds adopt unique structure types, with characteristic crystal-chemical features of the respective metal cations, namely a linear (or nearly) linear coordination of the Hg_2^{2+} and Hg^{2+} cations, a distorted octahedral coordination of the Cd^{2+} and Zn^{2+} cations, and a tetrahedral coordination of Cr in the oxochromate(VI) anions.

Conflicts of Interest: The authors declare no conflict of interest.

References and Notes

1. Faggiani, R.; Gillespie, R.J.; Vekris, J.E. The Cadmium(I) Ion, $(Cd_2)^{2+}$; X-ray Crystal Structure of $Cd_2(AlCl_4)_2$. *J. Chem. Soc. Chem. Commun.* **1986**, *7*, 517–518. [CrossRef]
2. Staffel, T.; Meyer, G. Synthesis and crystal structures of $Cd(AlCl_4)_2$ and $Cd_2(AlCl_4)_2$. *Z. Anorg. Allg. Chem.* **1987**, *548*, 45–54. [CrossRef]
3. Pyykkö, P. Relativistic effects in structural chemistry. *Chem. Rev.* **1988**, *88*, 563–594. [CrossRef]
4. Pyykkö, P. Relativistic Effects in Chemistry: More Common Than You Thought. *Annu. Rev. Phys. Chem.* **2012**, *63*, 45–64. [CrossRef] [PubMed]
5. Grdenić, D. The structural chemistry of mercury. *Quart. Rev. Chem. Soc.* **1965**, *19*, 303–328. [CrossRef]
6. Aurivillius, K. The structural chemistry of inorganic mercury(II) compounds. Some aspects of the determination of the positions of "light" atoms in the presence of "heavy" atoms in crystal structures. *Ark. Kemi* **1965**, *24*, 151–187.
7. Breitinger, D.K.; Brodersen, K. Development of and problems in the chemistry of mercury-nitrogen compounds. *Angew. Chem. Int. Ed. Eng.* **1970**, *5*, 357–367. [CrossRef]
8. Müller-Buschbaum, H. On the crystal chemistry of oxomercurates(II). *J. Alloys Compd.* **1995**, *229*, 107–122. [CrossRef]
9. Pervukhina, N.V.; Magarill, S.A.; Borisov, S.V.; Romanenko, G.V.; Pal'chik, N.A. Crystal chemistry of compounds containing mercury in low oxidation states. *Russ. Chem. Rev.* **1999**, *68*, 615–636. [CrossRef]
10. Borisov, S.V.; Magarill, S.A.; Pervukhina, N.V.; Peresypkina, E.V. Crystal chemistry of mercury oxo- and chalcohalides. *Cryst. Rev.* **2005**, *11*, 87–123. [CrossRef]
11. Breitinger, D.K. Cadmium and Mercury. In *Comprehensive Coordination Chemistry II*; McCleverty, J.A., Meyer, T.J., Eds.; Elsevier: Oxford, UK, 2004; pp. 1253–1292.
12. Archibald, S.J. Zinc. In *Comprehensive Coordination Chemistry II*; McCleverty, J.A., Meyer, T.J., Eds.; Elsevier: Oxford, UK, 2004; pp. 1147–1251.
13. Weil, M. Preparation and crystal structures of the isotypic compounds $CdXO_4 \cdot 2HgO$ (X = S, Se). *Z. Naturforsch.* **2004**, *59b*, 281–285. [CrossRef]
14. Weil, M. Preparation and crystal structure analyses of compounds in the systems $HgO/MXO_4/H_2O$ (M = Co, Zn, Cd; X = S, Se). *Z. Anorg. Allg. Chem.* **2004**, *630*, 921–927. [CrossRef]
15. Weil, M.; Stöger, B. Dimorphism im mercurous chromate—The crystal structures of α- and β-Hg_2CrO_4. *Z. Anorg. Allg. Chem.* **2006**, *632*, 2131. [CrossRef]
16. Weil, M.; Stöger, B. The mercury chromates $Hg_6Cr_2O_9$ and $Hg_6Cr_2O_{10}$—Preparation and crystal structures, and thermal behaviour of $Hg_6Cr_2O_9$. *J. Solid State Chem.* **2006**, *179*, 2479–2486. [CrossRef]
17. Stålhandske, C. Mercury(II) chromate. *Acta Crystallogr.* **1978**, *B34*, 1968–1969. [CrossRef]
18. Stöger, B.; Weil, M. Hydrothermal crystal growth and crystal structures of the mercury(II) chromates(VI) α-$HgCrO_4$, β-$HgCrO_4$ and $HgCrO_4(H_2O)$. *Z. Naturforsch.* **2006**, *61*, 708–714. [CrossRef]
19. Hansen, T.; Müller-Buschbaum, H.; Walz, L. Einkristallröntgenstrukturanalyse an Quecksilberchromat(VI): $Hg_3O_2CrO_4$. *Z. Naturforsch.* **1995**, *50*, 47–50.

20. Weil, M.; Stöger, B.; Zobetz, E.; Baran, E.J. Crystal structure and characterisation of mercury(II) dichromate(VI). *Monatsh. Chem.* **2006**, *137*, 987–996. [CrossRef]
21. Aurivillius, K.; Stålhandske, C. Neutron diffraction study of mercury(II) chromate hemihydrate, $HgCrO_4(H_2O)_{0.5}$. *Z. Kristallogr.* **1975**, *142*, 129–141.
22. Groat, L.A.; Roberts, A.C.; le Page, Y. The crystal structure of wattersite, $Hg_4^{1+}Hg^{2+}Cr^{6+}O_6$. *Can. Mineral.* **1995**, *33*, 41–46.
23. Klein, W.; Curda, J.; Friese, K.; Jansen, M. Dilead mercury chromate(VI), Pb_2HgCrO_6. *Acta Crystallogr.* **2002**, *C58*, i23–i24. [CrossRef]
24. Klein, W.; Curda, J.; Jansen, M. Dilead trimercury chromate(VI), $Pb_2(Hg_3O_4)(CrO_4)$. *Acta Crystallogr.* **2005**, *C61*, i63–i64.
25. Brandt, K. X-ray Analysis of $CrVO_4$ and Isomorphous Compounds. *Ark. Kemi Mineral. Geol.* **1943**, *17*, 1–13.
26. Riou, A.; Lecerf, A. Les hydroxychromates $M_2(OH)_2CrO_4$ (M = Mg^{2+}, Ni^{2+}, Zn^{2+}). *C. R. Acad. Sci. Paris* **1970**, *C270*, 1109–1112.
27. Muller, O.; White, W.B.; Roy, R. X-ray diffraction study of the chromates of nickel, magnesium and cadmium. *Z. Kristallogr.* **1969**, *130*, 112–120. [CrossRef]
28. Weil, M. The crystal structures of $Hg_7Se_3O_{13}H_2$ and $Hg_8Se_4O_{17}H_2$—Two mixed-valent mercury oxoselenium compounds with a multifarious crystal chemistry. *Z. Kristallogr.* **2004**, *219*, 621–629. [CrossRef]
29. Weil, M. The Mixed-valent Mercury(I/II) Compounds $Hg_3(HAsO_4)_2$ and $Hg_6As_2O_{10}$. *Z. Naturforsch.* **2014**, *69*, 665–673. [CrossRef]
30. Weil, M.; Tillmanns, E.; Pushcharovsky, D.Y. Hydrothermal Single-Crystal Growth in the Systems Ag/Hg/X/O (X = V^V, As^V): Crystal Structures of $(Ag_3Hg)VO_4$, $(Ag_2Hg_2)3(VO_4)_4$, and $(Ag_2Hg_2)_2(HgO_2)(AsO_4)_2$ with the Unusual Tetrahedral Cluster Cations $(Ag_3Hg)^{3+}$ and $(Ag_2Hg_2)^{4+}$ and Crystal Structure of $AgHgVO_4$. *Inorg. Chem.* **2005**, *44*, 1443–1451. [CrossRef] [PubMed]
31. Pressprich, M.R.; Willett, R.D.; Poshusta, R.D.; Saunders, S.C.; Davis, H.B.; Gard, H.B. Preparation and crystal structure of dipyrazinium trichromate and bond length correlation for chromate anions of the form $Cr_nO_{3n+1}^{2-}$. *Inorg. Chem.* **1988**, *27*, 260–264. [CrossRef]
32. Siidra, O.I.; Krivovichev, S.V.; Filatov, S.K. Minerals and synthetic Pb(II) compounds with oxocentered tetrahedra: Review and classification. *Z. Kristallogr.* **2008**, *223*, 114–125. [CrossRef]
33. Krivovichev, S.V.; Mentre, O.; Siidra, O.I.; Colmont, M.; Filatov, S.K. Anion-centered tetrahedra in inorganic compounds. *Chem. Rev.* **2013**, *113*, 6459–6535. [CrossRef] [PubMed]
34. Addison, A.W.; Nageswara Rao, T.; Reedijk, J.; van Rijn, J.; Verschoor, G.C. Synthesis, structure, and spectroscopic properties of copper(II) compounds containing nitrogen–sulphur donor ligands; the crystal and molecular structure of aqua[1,7-bis(*N*-methylbenzimidazol-2′-yl)-2,6-dithiaheptane]copper(II) perchlorate. *J. Chem. Soc. Dalton Trans.* **1984**, 1349–1356. [CrossRef]
35. Extreme values of τ_5 for a five-coordinate atom are 0 for a square-pyramidal arrangement and 1 for a trigonal-bipyramidal arrangement.
36. Brown, I.D. *The Chemical Bond in Inorganic Chemistry: The Bond Valence Model*. Oxford University Press: Oxford, UK, 2002.
37. Brese, N.E.; O'Keeffe, M. Bond-valence parameters for solids. *Acta Crystallogr.* **1991**, *B47*, 192–197. [CrossRef]
38. Values of GII < 0.10 v.u. suggest that little or no strain is present in the crystal structure while values between 0.10 and 0.20 v.u. indicate a significant lattice-induced strain. Values > 0.20 v.u. point to an instability of the structure due to too much strain.
39. Herrendorf, W.H. *Program for Optimization of the Crystal Shape for Numerical Absorption Correction*; Universities of Karlsruhe: Gießen, Germany, 1997.
40. Sheldrick, G.M. Phase annealing in *SHELX*-90: Direct methods for larger structures. *Acta Cryst.* **1990**, *A46*, 467–473. [CrossRef]
41. Sheldrick, G.M. Crystal structure refinement with *SHELXL*. *Acta Cryst.* **2008**, *A64*, 112–122. [CrossRef] [PubMed]
42. Dowty, E. *ATOMS*; Shape Software: Kingsport, TN, USA, 2006.

crystals

MDPI

Article

Stuck in Our Teeth? Crystal Structure of a New Copper Amalgam, Cu$_3$Hg

Jonathan Sappl, Ralph Freund and Constantin Hoch *

Department Chemie, LMU München, Butenandtstraße 5-13(D), D-81377 München, Germany; jonathan.sappl@cup.uni-muenchen.de (J.S.); ralph.freund@campus.lmu.de (R.F.)
* Correspondence: constantin.hoch@cup.uni-muenchen.de; Tel.: +49-89-2180-77421

Academic Editor: Matthias Weil
Received: 23 October 2017; Accepted: 22 November 2017; Published: 24 November 2017

Abstract: We have synthesized a new Cu amalgam, the Cu-rich phase Cu$_3$Hg. It crystallizes with the Ni$_3$Sn structure type with a hexagonal unit cell (space group $P6_3/mmc$, $a = 5.408(4)$ Å, $c = 4.390(3)$ Å) and shows some mixed occupancy of Cu on the Hg site, resulting in a refined composition of Cu$_{3.11}$Hg$_{0.89}$. This is the first example of an amalgam with the Ni$_3$Sn structure type where Hg is located mainly on the Sn site. Cu$_3$Hg might be one of the phases constituting dental amalgams and therefore has major relevance, as well as the only Cu amalgam phase described so far, Cu$_7$Hg$_6$ with the γ-brass structure. It occurs as a biphase in our samples. Thermal decomposition of Cu amalgam samples in a dynamic vacuum yields nanostructured copper networks, possibly suitable for catalytic applications.

Keywords: copper amalgams; dental amalgams; crystal structure; Ni$_3$Sn structure type

1. Introduction

Modern dental amalgams consist of Hg, Ag, Sn, Cu and minor metallic additives. In order to prepare a typical amalgam filling, liquid mercury is mixed with powders or shavings of an alloy containing Ag, Sn and Cu in the suitable composition: 40–70 weight % Ag, 12–30% Sn and 12–24% Cu. Additives can be Zn, In or Pd up to ca. 4% [1–3]. The alloy is mixed with an equal weight amount of Hg to form the plastic amalgam, solidifying within minutes to hours. The solidification goes along with the formation of a number of intermetallic phases. The most important in the resulting metal matrix composite are Ag$_2$Hg$_3$ [4] as the major matrix phase, Ag$_3$Sn [4,5] as the mechanically strongest phase, Sn$_8$Hg [6] as the most corrosive phase, Ag$_5$Sn [7], Cu$_6$Sn$_5$ and Cu$_3$Sn [8–10], amongst others [11,12]. The γ2-phase Sn$_8$Hg is posing a major problem as it is relatively corrosive and leads to degradation of the filling over the years, especially at the tooth-amalgam interface. Today, so-called γ2-free amalgams are in use, which are higher in Cu content [13,14]. As the reaction takes place at physiological temperatures, the resulting alloy may not be a thermodynamically stable system. Other combinations of the metal atoms may also be considered: Cu amalgams and Ag-Cu solid solutions, as well as ternary and quaternary phases may also come into play at equilibrium conditions.

First reports on the employment of amalgams in dental fillings date back to 659 during the Tang dynasty in ancient China [15]; in European medical history, the first evidence of amalgam fillings adopting the Chinese recipe dates back to the end of the 16th century [16]. Despite the slow, but constant replacement of amalgams by other dental filling materials [17], many people still rely on amalgam fillings as they are resistive for a very long time, and the mechanical removal of existing amalgam fillings harbors the risk of unnecessary Hg release. With respect to the vast employment of dental amalgams, it is very surprising that new binary Cu-Hg phases can be found. It becomes even close to improbable when taking into account that both mercury and copper have been important materials since prehistoric times: the combination of both is employed, e.g., in the centuries-old fire gilding process where a copper surface is amalgamated by submerging in a Hg nitrate solution prior to applying gold amalgam and subsequent distillation of the Hg. Despite the widespread and longtime preparation of Cu-Hg phases, knowledge hereon is scarce. The published phase diagram [18] shows only one phase, Cu_7Hg_6, and even for this phase, there exist concurring structure descriptions, some of them expressing reasonable doubt about the others [19–22]. Details on the behavior of the liquidus curve or thermodynamic stability ranges also remain unclear.

The problem with identifying amalgam phases often lies in their low thermal stability. As Hg has a very low melting point, the liquidus curve in most phase diagrams shows a steep run in the Hg-rich region, and the Hg-richest phases normally show low peritectic decomposition temperatures (see Table 1). Standard solid state preparation techniques usually are not suitable to prepare those delicate phases. Thermoanalytical methods can show the presence of low-decomposing phases; however, if accompanying structural elucidations are impeded, their exact compositions remain unclear. In addition, thermoanalytical studies can be hampered by kinetic effects, which are much more pronounced at low than at high temperatures (seed formation, phase transformations and the like). Problems with structural analyses of crystalline amalgams with high Hg content are caused by their high X-ray absorption coefficients. A special case is given for the widely-employed Au amalgams, which show an especially inconvenient combination of high absorption and especially low X-ray contrast, leading to only very few structurally-sufficiently described phases and a rudimentary phase diagram [18].

Table 1. Decomposition temperatures or melting points of Hg-rich amalgams AHg_x with $x > 3$ and with A = non-Hg metal. Values are taken from [18] if not given otherwise. Congruent melting temperatures are given in italic numbers.

Composition	Hg/A Ratio x	Decomposition Temperature (°C)
Cs_2Hg_{27}	13.5	12 [23]
KHg_{11}	11.0	70 [24]
$RbHg_{11}$	11.0	70 [24]
$CaHg_{11}$	11.0	84
$SrHg_{11}$	11.0	63 [24]
$BaHg_{11}$	11.0	162 [24]
$(N(CH_3)_4)Hg_8$	8.0	9 [25]
Rb_3Hg_{20}	6.67	132
Cs_3Hg_{20}	6.67	*158*
KHg_6	6.0	170 [26]
$BaHg_6$	6.0	*410*
$Ba_{20}Hg_{103}$	5.15	505
K_7Hg_{31}	4.43	187
Rb_7Hg_{31}	4.43	162
$A_{11}Hg_{45}$ [1]	4.09	145–720
$PtHg_4$	4.0	≥ 200 [27]
Rb_5Hg_{19}	3.8	193
Cs_5Hg_{19}	3.8	*164*
K_3Hg_{11}	3.67	195
$A_{14}Hg_{51}$ [2]	3.64	157–480
K_2Hg_7	3.5	202 [28]
Rb_2Hg_7	3.5	197 [28]

[1] Amalgams with the $Sm_{11+x}Cd_{45-x}$ structure type show compositional ranges between $x = 3.79$ (La, $T = 720$ °C) and 4.09 (Gd, $T = 145$ °C) and have been reported for A = La, Ce, Pr, Nd, Sm, Gd and U [29,30]. [2] Amalgams with variants of the $Gd_{14+x}Ag_{51-x}$ structure type show compositional ranges between $x = 4.63$ (Na, $T = 157$ °C) and 5.44 (Sr, $T = 480$ °C) and have been described for A = Na, Ca, Sr, Eu and Yb [29,31–33].

It now becomes clear that with all the named complications, a number of new amalgams can be expected to be found even in systems that have been employed technically for a very long time. Both synthesis and structural analysis require non-standard techniques and diligence.

2. Results

Single crystals of Cu_3Hg were found together with crystals of Cu_7Hg_6 on the surface of an amalgamated copper spoon after 12 days at 90 °C. Both phases form crystals of bronze color when Hg-free and of bright silver luster when a thin film of Hg covers the surface. Cu_3Hg forms prismatic crystals, whereas Cu_7Hg_6 forms platelets. Both amalgams are air-stable at room temperature at least over several weeks. A rod-shaped crystal of Cu_3Hg was selected under a binocular and glued on top of a glass fiber. Data were carefully corrected for considerable absorption effects on the basis of indexed crystal faces. Metric and extinction conditions pointed towards the Ni_3Sn structure type (space group $P6_3/mmc$), and structure solution and refinement showed this to be correct. Crystallographic details and results of the single-crystal structure refinement are compiled in Tables 2–5.

Table 2. Crystallographic data and details on single crystal data collection, structure solution and refinement for $Cu_{3.11(4)}Hg_{0.89(4)}$. Data collection was performed at room temperature. All standard deviations are given in parentheses in units of the last digit.

Composition	$Cu_{3.11(4)}Hg_{0.89(4)}$
Crystal system	hexagonal
Space group	$P6_3/mmc$, (No. 194)
Lattice parameters, a, c (Å)	5.408(4), 4.390(3)
Unit cell volume, V (Å^3)	111.18(13)
No. of formula units, Z	2
Density (X-ray), ρ (g cm^{-3})	11.235
Diffractometer	STOE IPDS 1, AgK$_\alpha$ radiation
Data collection temperature, T (K)	295(2)
Absorption coefficient, μ (mm^{-1})	48.62
Diffraction angle range, ϑ (°)	5.02–23.53
Index range	$-7 \leq h, k \leq 7, -6 \leq l \leq 6$
No. of collected reflections	2154
No. of independent reflections	72
No. of independent reflections ($I \geq 2\sigma(I)$)	71
R_{int}	0.2198
R_σ	0.0485
Structure factor, $F(000)$	322.8
Corrections	Lorentz, polarization, absorption effects
Absorption correction	numerical [34,35]
Structure solution	direct methods [36]
Structure refinement	full-matrix least-squares on F^2 [36]
No. of lest-squares parameters	9
GooF	1.180
R values ($I \geq 2\sigma(I)$)	$R1 = 0.0345, wR2 = 0.0757$
R values (all data)	$R1 = 0.0368, wR2 = 0.0773$
Residual $\rho(e^-)$ max/min (e$^-$Å^{-3})	+1.698/−1.419
Extinction coefficient	0.047(4)

Cu_3Hg crystallizes with the Ni_3Sn structure type. The lattice parameter a of the Ni_3Sn structure type is doubled with respect to simple hcp by creating two crystallographic sites with non-equal Wyckoff numbers for Sn (Hg) on $2d$ and Ni (Cu) on $6h$ while retaining the original space group type $P6_3/mmc$. This shows the Ni_3Sn structure type to be a coloring variant of hcp as the hcp topology is retained in an ordered packing of two different spheres. By the coloring, Hg atoms are only coordinated by 12 Cu atoms forming [HgCu$_{12}$] anticuboctahedra. Cu atoms form streaks of *trans*-face-sharing [Cu$_6$] octahedra (see Figure 1). In an alternative picture, the crystal structure can be visualized with plane hexagonal nets at heights $z = 1/4$ and $3/4$, which are stacked along c with ...ABAB... periodicity, where A and B are shifted by (1/3, 2/3, 1/2) with respect to each other.

During structure refinement, unusual residual electron densities and weight factors pointed towards possible mixed occupation on the two crystallographic sites. This was tested by independently refining mixed occupations with retaining an overall full occupation of the two sites. The Cu2 site on $6h$ shows no sign of mixed occupation by Hg, whereas there is significant Cu content on the Hg/Cu1 site on $2d$ (see Table 3). Refinement of the mixed occupation according to the denotation $Cu_3(Hg_{1-x}Cu_x)$ leads to $x = 0.11(4)$ and the overall refined composition $Cu_{3.11(4)}Hg_{0.89(4)}$.

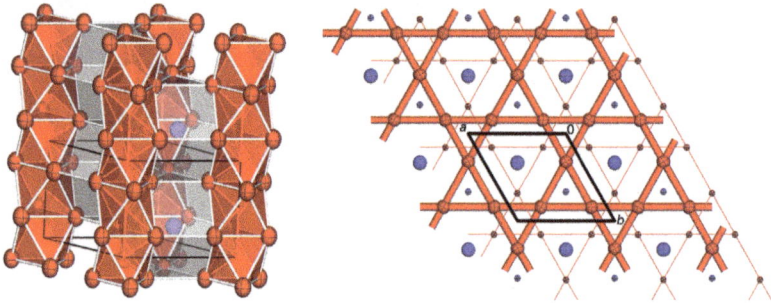

Figure 1. Crystal structure of Cu_3Hg with Ni_3Sn structure type. **Left**: Polyhedra packing with blue Hg atoms centering light grey $[HgCu_{12}]$ anticuboctahedra and red Cu atoms forming empty *trans*-face-sharing $[Cu_6]$ octahedra, c pointing upwards. **Right**: Net representation in the ab plane at height $z = 1/4$ (fat bonds, large atoms) and $z = 3/4$ (thin bonds, small atoms) with the unit cell given in black. Stacking of the nets follows periodicity ...ABAB... along c while shifting B versus A by $(1/3, 2/3, 1/2)$ as shown. Ellipsoids are drawn at a probability level of 99% for Cu (red) and Hg (blue).

Figure 2 shows a crystal of Cu_3Hg under an optical microscope (left) and the specimen used for X-ray data collection after EDX measurements (right). The severe degradation of Cu_3Hg in an electron beam under high vacuum becomes evident. The sensitivity of the amalgam towards decomposition in a vacuum hampers detailed EDX analyses; still, we managed to record spectra at three different points of the crystal. The recorded spectra show the absence of further metal atoms and a composition similar to the single-crystal refinement results, but slightly lower in Cu content. The mean measured composition is $Cu_{2.79}Hg$ (see Table 6), calculated: $Cu_{3.49}Hg$. This is somehow surprising as decomposition in a high vacuum of the electron microscope can clearly be seen in the crystal picture, which would mean evaporation of Hg and therefore a higher Cu content with respect to the results from single-crystal refinement. A reason for this deviation could be the formation of a Hg surface layer after decomposition and prior to evaporation of the Hg in the vacuum. As EDX primarily is a surface-sensitive method, Hg may thus be overrepresented in the results.

Figure 2. Left: Photograph of a Cu_3Hg single crystal under an optical microscope (size of the crystal: 87×13 µm) showing well-shaped faces, the typical prismatic habitus and bronze luster. **Right**: Scanning electron micrograph of a Cu_3Hg crystal after EDX measurements. Deterioration effects due to high vacuum and electron beam are clearly visible.

In order to further explore the phase width of $Cu_{3+x}Hg_{1-x}$ and to gather knowledge about thermal stability and phase relations of both Cu amalgams Cu_3Hg and Cu_7Hg_6, thermochemical investigations are in progress.

Table 3. Standardized fractional atomic coordinates [37] and equivalent isotropic displacement parameters ($Å^2$) for $Cu_{3.11(4)}Hg_{0.89(4)}$. The equivalent isotropic displacement parameter is defined as 1/3 of the trace of the anisotropic displacement tensor. All standard deviations are given in parentheses in units of the last digit.

Atom	Occupation Factor	Wyckoff Letter	x	y	z	U_{equiv}
Hg1	0.89(5)	2d	$\frac{1}{3}$	$\frac{2}{3}$	$\frac{3}{4}$	0.0160(7)
Cu1	0.11(5)	2d	$\frac{1}{3}$	$\frac{2}{3}$	$\frac{3}{4}$	0.0160(7)
Cu2	1	6h	0.1589(3)	$2x$	$\frac{1}{4}$	0.0201(18)

Table 4. Coefficients U_{ij} of the anisotropic displacement tensor ($Å^2$) for $Cu_{3.11(4)}Hg_{0.89(4)}$. U_{ij} is defined as $U_{ij} = \exp\{-2\pi^2[U_{11}(ha^*)^2 + ... + 2U_{21}hka^*b^*]\}$. All standard deviations are given in parentheses in units of the last digit.

Atom	U_{11}	U_{12}	U_{33}	U_{23}	U_{13}	U_{12}
Hg1	0.0156(7)	$=U_{11}$	0.0168(8)	0	0	0.0078(4)
Cu1	0.0156(7)	$=U_{11}$	0.0168(8)	0	0	0.0078(4)
Cu2	0.0195(19)	0.024(2)	0.019(2)	0	0	0.0118(10)

Table 5. Selected interatomic distances and their frequencies in $Cu_{3.11(4)}Hg_{0.89(4)}$ in Å. All standard deviations are given in parentheses in units of the last digit.

Atom 1	Atom 2	Distance
Hg1	Cu2	2.7048(19) (6×)
	Cu2	2.736(2) (6×)
Cu2	Cu2	2.578(5) (2×)
	Cu2	2.6521(19) (4×)

Table 6. Results of EDX spectroscopic measurements on the Cu_3Hg crystal shown in Figure 2 (right). The values are averaged over three measurements on different spots. The largest deviations from the mean value are given in parentheses. Expected values are calculated for the composition $Cu_{3.49}Hg$.

Atom	Atom-% Detected	Atom-% Calculated
Cu	73.6(2)	77.7
Hg	26.4(2)	22.3

3. Discussion

All amalgams adopting the Ni_3Sn structure type [38–41] are compiled in Table 7. Always, the minority component *A* is coordinated only by Hg atoms in an anticuboctahedral environment, and all known examples have the Hg-rich composition MHg_3. This can be seen as an expression of some Coulombic contributions $A^{\delta+}[Hg_3]^{\delta-}$ and the formation of coordination spheres of negatively polarized Hg atoms around the cationic species *A*. This is corroborated by the fact that this class of amalgams is only formed by explicitly electropositive *A* metals.

The Ni_3Sn structure type is a favored sphere packing in amalgams where the atomic radii of *A* do not differ much from the one of Hg (151 pm, [42]). Differences according to Table 7 seem only favorable if the A metal is larger than Hg. The largest difference can be found for Sr with an atomic radius 42% larger than the one of Hg. No amalgams are known with A atoms smaller than Hg so far. In addition, it can be stated that all known amalgams of this type show a ratio of the lattice parameters c/a smaller than its ideal value for a packing of incompressible spheres of equal size with all spheres in contact. The ab plane is widened with respect to the stacking of the planes. This distortion cannot be due to only geometrical reasons because the larger A atoms should distort the packing equally in all directions, not only the ab plane. The anticuboctahedral coordination polyhedra $[AHg_{12}]$ thus are oblate with the six equatorial Hg atoms having larger distances than the other six. The variation of the c/a ratio does not follow the quotient of the atomic radii: c/a is largest (and therefore closest to the ideal geometric ratio) for LiHg$_3$ (0.768) and for YbHg$_3$ (0.761), while Li and Hg have almost identical atomic radii, and Yb is 17% larger than Hg. The smallest c/a ratios are found for LaHg$_3$ (0.729, La being 24% larger than Hg), CeHg$_3$ (0.731, 21%), Th (0.730, 19%) and U (0.735, 3%). Geometric, as well as electronic reasons may contribute to the structural distortion, and heteroatomic interactions seem to have higher bonding contributions than homoatomic ones, as suggested by studies on the electronic structure of Ni$_3$Sn [39]. In the amalgam cases, Coulombic contributions may be the underlying reason; however, the degree of interplay of different contributions still remains speculative. One could assume bonding Hg$^{\delta-}$-Hg$^{\delta-}$ interactions to cause a contraction of the $[AHg_{12}]$ anticuboctahedra along the [001] direction, as Hg$^{\delta-}$ formally behaves like an early p group metal [43] and tends to form Hg-Hg bonding.

A similar case of deviation from ideal sphere packing geometry is known for elemental mercury. Rhombohedral α-Hg crystallizes in a distortion variant of fcc where the individual hexagonal sphere planes are widened with respect to their stacking along the {111} directions of the cubic fcc cell, leading to a rhombohedral metric with $\alpha = 70.44(6)°$ [44] instead of 60° for an ideal fcc packing. Band structure calculations show differences in bond strengths for σ- and π-type interactions between the Hg(p_z) orbitals within and between the hexagonal layers, resulting in the structural distortion of the sphere packing [45,46].

Table 7. Amalgams AHg_3 (A = non-Hg metal) adopting the Ni$_3$Hg structure type. Hg has an atomic radius of 151 pm [42]. The ideal c/a ratio for hcp is 1.633. Due to a doubling of a, the ideal c/a ratio in the Ni$_3$Sn structure type is 0.8167. In Ni$_3$Sn (r_{Ni} = 124 pm, r_{Sn} = 132 pm, ratio r_{Sn}/r_{Ni} = 1.06), the c/a ratio is 0.802 [40].

Phase AHg_3	r_A [pm]	r_A/r_{Hg}	c/a
LiHg$_3$ [29,47]	152	1.01	0.768
CaHg$_3$ [31]	197	1.30	0.757
SrHg$_3$ [31,48]	215	1.42	0.741
ScHg$_3$ [29,49]	160	1.06	0.748
YHg$_3$ [29,49,50]	181	1.20	0.744
LaHg$_3$ [29,51]	187	1.24	0.729
CeHg$_3$ [52]	182	1.21	0.731
EuHg$_3$ [31]	180	1.19	0.747
GdHg$_3$ [29,53]	180	1.19	0.742
TbHg$_3$ [53]	177	1.17	0.748
DyHg$_3$ [29,53]	178	1.18	0.746
HoHg$_3$ [29,53]	176	1.17	0.747
ErHg$_3$ [29,53]	176	1.17	0.748
TmHg$_3$ [29,53]	176	1.17	0.748
YbHg$_3$ [53]	176	1.17	0.761
LuHg$_3$ [29,53]	174	1.15	0.750
ThHg$_3$ [54]	179	1.19	0.730
UHg$_3$ [55]	156	1.03	0.735

Most interesting within this context is Cu$_3$Hg as the first amalgam in this structure type with a reversed composition. In comparison to the other amalgams, Cu$_3$Hg may be seen as the *anti*-Ni$_3$Sn-type, as it is the only amalgam where the A element constitutes the majority component, and Hg is coordinated in an anticuboctahedron of Cu atoms. The structural deviation from ideal geometry leads to a c/a ratio of 0.812, which is much closer to the ideal value of 0.8167 than for all amalgams with AHg$_3$ composition. Here, the bonding situation is obviously quite different. Obviously, Coulombic interactions cannot play a major role here, as Cu is not a very electropositive metal. This would explain the absence of bonding Hg-Hg contributions and therefore no contraction along c.

While the details of chemical bonding still remain unclear, the inversed composition may also be explained by geometrical reasons. If the Ni$_3$Hg structure type demands a structural necessity for the minority component to be the bigger atom, then Cu with atomic radius of 128 pm [42] favors the formation of Cu$_3$Hg with a quotient r_{Hg}/r_{Cu} = 1.18 over the formation of CuHg$_3$ with a quotient of r_{Cu}/r_{Hg} = 0.878. If the atomic radii really play a decisive role in the formation of amalgams of the Ni$_3$Sn structure type with a composition A_3Hg, then the respective amalgams should also be found for A = Cr (r = 128 pm), Mn (r = 127 pm), Fe (r = 126 pm), Co (r = 125 pm) and Ni (r = 124 pm). As from these elements only Mn and Ni are known to form amalgams and as from these no amalgam with composition A_3Hg is known so far, we consider it worthwhile to have a closer look at the respective binary phase diagrams in the near future.

4. Materials and Methods

4.1. Synthesis of Copper Amalgams

We did not intend to prepare copper amalgams. Crystals of Cu$_3$Hg and Cu$_7$Hg$_6$ were the serendipitous results of the attempt to synthesize Hg-rich uranium amalgams by electro-crystallization, starting from a solution of UI$_4$ in *N,N*-dimethylformamide (DMF) and a reactive cathode consisting of a Hg drop suspended in an amalgamated Cu spoon. The preparative approach to temperature-sensitive Hg-rich amalgams via electro-crystallization had proven very convenient in previous cases [23,32,33,56,57]. Electro-crystallization was performed at 90 °C for two weeks and resulted in the formation of nanocrystalline UO$_2$ (see Figure 3, left), while the Hg drop formed the copper amalgams Cu$_3$Hg and Cu$_6$Hg$_7$ over the reaction time of two weeks. The reason for the unsuccessful U amalgam synthesis is that UI$_4$ dissolves in molecular form in DMF and does not form ionic complexes [U(DMF)$_x$]$^{4+}$ as other metal iodides do. This has been shown by recording UV-Vis spectra of the solutions and comparison with the literature.

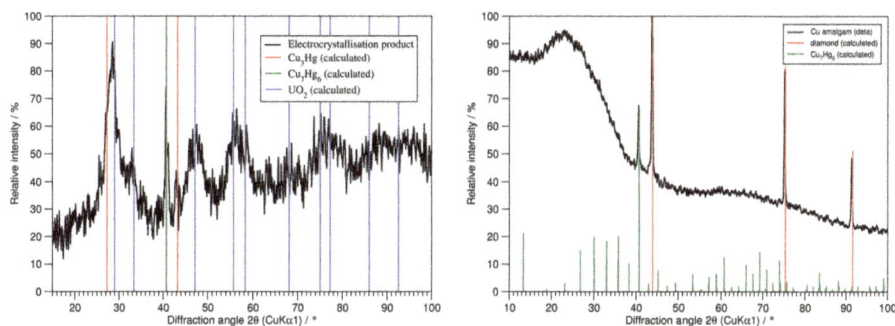

Figure 3. **Left**: Powder diffraction pattern of the electro-crystallization product. The broad maxima originate from nano-scaled UO$_2$; the sharp maxima belong to both Cu amalgams Cu$_3$Hg and Cu$_7$Hg$_6$. **Right**: Powder diffraction pattern of a sample of Cu amalgams. Data collection was performed in parafocusing Debye–Scherrer transmission geometry on a capillary sample diluted with diamond powder.

The reaction of a copper foil with mercury at 105 °C was repeated outside of the electro-crystallization chamber and also resulted in the formation of the Cu amalgams Cu$_3$Hg and Cu$_7$Hg$_6$ directly from the elements; see Figure 3. Both amalgams decompose at temperatures between

110 and 200 °C. By heating the amalgamated copper foil in a dynamic vacuum to 200 °C, the released mercury is distilled off, and a homogeneous copper network with interpenetrating nano-scaled copper rod structures is formed (see Figure 4). This copper network with high porosity might be interesting for utilization in heterogeneous catalysis reactions due to its simple and straightforward preparation.

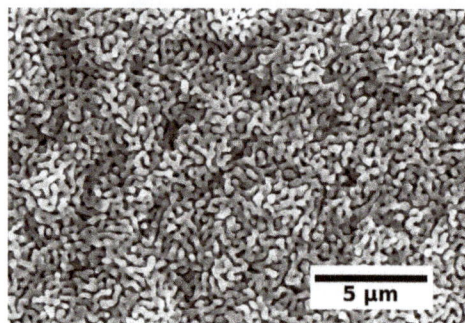

Figure 4. Scanning electron micrograph of nano-scaled metallic copper structures, prepared by thermolysis of Cu amalgams at 200 °C in a vacuum.

4.2. Single Crystal Investigations

Single crystals of the copper amalgams were selected under binoculars, and specimens with well-shaped faces (see Figure 2, left) were glued on top of a glass fiber with cyanoacrylate glue. They were centered on the one-circle goniometer of a diffractometer system IPDS 1 (Stoe & Cie., Darmstadt, Germany) equipped with graphite-monochromatized Ag-Kα radiation (fine-focus X-ray tube) and an imaging plate detector After checking crystal quality and crystal metrics, data collection was performed in φ scans; data of the accessible part of an entire Ewald sphere were collected. Data were corrected for Lorentz and polarization effects with the diffractometer software package [58]. Absorption corrections were performed carefully on the basis of optimized indexed crystal faces [34,35,58]. Structure solution for Cu_3Hg was performed with direct methods [36] in space group $P6_3/mmc$ as indicated by extinction conditions together with statistics on E^2-1 (E being the normalized structure factor). Structure refinement was performed with full-matrix least-squares cycles [36] on F^2 (F being the structure factor). Both atoms were treated with anisotropic displacement tensors. Further information can be obtained from Fachinformationszentrum Karlsruhe, 76344 Eggenstein-Leopoldshafen, Germany (fax +49 (0)7247 808 666; e-mail: crysdata@fiz-karlsruhe.de; http://www.fiz-informationsdienste.de/en/DB/icsd/depot_anforderung.html), on quoting the deposition number CSD-433425.

4.3. Powder Diffractometry

For powder diffraction, the amalgam samples were prepared by grinding together with diamond powder for both optical dilution due to high absorption coefficients and mechanical reasons as the amalgams are rather ductile. The powders were sealed in glass capillaries ($\varnothing$ = 0.3 mm), and data collection was performed on a diffractometer system STADI P (Stoe & Cie., Darmstadt, Germany) equipped with Ge-monochromatized MoKα1 radiation and a Mythen 2K detector in parafocusing Debye–Scherrer geometry. For better comparison of the recorded diffraction patterns with calculated patterns based on single-crystal data from literature, they were converted to Cu-Kα1 wavelength.

4.4. EDX Spectroscopy

EDX spectra were collected on an electron microscope system JSM-6500F (Jeol, Freising, Germany) with a field emission source and EDX detector at 25 kV.

Acknowledgments: We thank Wolfgang Schnick from Ludwig-Maximilians-Universität München for generous funding. We also thank Christian Minke from LMU München for EDX measurements and SEM imaging. Florian Kraus from Philipps-Universität Marburg kindly provided samples of UI_4 for the electro-crystallization experiments.

Author Contributions: C.H. conceptualized the studies and wrote the manuscript. J.S. supervised the practical work, prepared the single-crystal samples and performed all analyses. R.F. performed the syntheses.

Conflicts of Interest: The authors declare no conflict of interest.

References

1. Bharti, R.; Wadhwani, K.K.; Tikku, A.P.; Chandra, A. Dental amalgam: An update. *J. Conserv. Dent.* **2010**, *13*, 204–208, doi:10.4103/0972-0707.73380.

2. Sakaguchi, R.L.; Powers, J.M. *Craig's Restorative Dental Materials*, 13rd ed.; Mosby: St. Louis, MI, USA, 2011; ISBN 978-0323081085.

3. Bonsor, S.; Pearson, G. *A Clinical Guide to Applied Dental Materials*, 1st ed.; Churchill Livingstone: London, UK, 2012; ISBN 978-0702031588.

4. Fairhurst, C.W.; Cohen, J.B. The crystal structures of two compounds found in dental amalgam: Ag_2Hg_3 and Ag_3Sn. *Acta Crystallogr.* **1972**, *B28*, 371–378, doi:10.1107/S0567740872002432.

5. Rossi, P.J.; Zotov, N.; Mittemeijer, E.J. Redetermination of the crystal structure of the Ag_3Sn intermetallic compound. *Z. Kristallogr.* **2016**, *231*, 1–9, doi:10.1515/zkri-2015-1867.

6. Che, G.C.; Ellner, M.; Schubert, K. The hP1-type phases in alloys of cadmium, mercury, and indium with tin. *J. Mater. Sci.* **1991**, *26*, 2417–2420, doi:10.1007/BF01130189.

7. King, H.W.; Massalski, T.B. Lattice spacing relationships and the electronic structure of h.c.p. ζ-phases based on silver. *Philos. Mag.* **1961**, *6*, 669–682, doi:10.1080/14786436108244417.

8. Larsson, A.K.; Stenberg, L.; Lidin, S. The superstructure of the domain-twinned η'-Cu_6Sn_5. *Acta Crystallogr.* **1994**, *B50*, 636–643, doi:10.1107/S0108768194004052.

9. Watanabe, Y.; Fujinaga, Y.; Iwasaki, H. Lattice modulation in the long-period superstructure of Cu_3Sn. *Acta Crystallogr.* **1983**, *B39*, 306–311, doi:10.1107/S0108768183002451.

10. Xiahan, S.; Kui, D.; Hengqiang, Y. An ordered structure of Cu_3Sn in Cu-Sn alloy investigated by transmission electron microscopy. *J. Alloys Compd.* **2009**, *469*, 129–136, doi:10.1016/j.jallcom.2008.01.107.

11. Mitchell, R.J.; Okabe, T. Setting reactions in dental amalgam. Part 1: Phases and microstructures between one hour and one week. *Crit. Rev. Oral Biol. Med.* **1996**, *7*, 12–22, doi:10.1177/10454411960070010101.

12. Mitchell, R.J.; Okabe, T. Setting reactions in dental amalgam. Part 2: The kinetics of amalgamation. *Crit. Rev. Oral Biol. Med.* **1996**, *7*, 23–25, doi:10.1177/10454411960070010201.

13. Beech, D.R. High copper alloys for dental amalgam. *Int. Dent. J.* **1982**, *32*, 240–251.

14. Städtler, P. Dental amalgam. I: Conventional and non-gamma-2 amalgams. *Int. J. Clin. Pharmacol. Ther. Toxicol.* **1991**, *29*, 161–163.

15. Chu, H.-T. The use of amalgam as filling material in dentistry in ancient china. *Chin. Med. J.* **1958**, *76*, 553–555.

16. Czarnetzki, A.; Ehrhardt, S. Re-dating the chinese amalgam filling of teeth in Europe. *Int. J. Anthropol.* **1990**, *5*, 325–332.

17. Newman, S.M. Amalgam alternatives: What can compete? *J. Am. Dent. Assoc.* **1991**, *122*, 67–71, doi:10.14219/jada.archive.1991.0246.

18. Massalski, T.B.; Okamoto, H.; Subramanian, P.R. *Binary Alloy Phase Diagrams*; ASM International: Materials Park, OH, USA, 1990; ISBN 978-0871704030.

19. Lindahl, T.; Westman, S. The structure of the rhombohedral gamma brass like phase in the copper-mercury system. *Acta Chem. Scand.* **1969**, *23*, 1181–1190.

20. Bernhardt, H.J.; Schmetzer, K. Belendorffite, a new copper amalgam dimorphous with kolymite. *Neues Jahrb. Mineral. Monatsh.* **1992**, *1992*, 18–21.

21. Markova, E.A.; Chernitsova, N.M.; Borodaev, N.M.; Yu, S.; Dubakina, L.S.; Yushko-Zakharova, O.E. The new mineral kolymite, Cu_7Hg_6. *Int. Geol. Rev.* **1982**, *24*, 233–237, doi:10.1080/00206818209452397.

22. Carnasciali, M.M.; Costa, G.A. Cu_xHg_y: A puzzling compound. *J. Alloys Compd.* **2001**, *317–318*, 491–496, doi:10.1016/S0925-8388(00)01376-1.

23. Hoch, C.; Simon, A. Cs$_2$Hg$_{27}$, the mercury-richest amalgam with close relationship to the Bergman phases. *Z. Anorg. Allg. Chem.* **2008**, *634*, 853–856, doi:10.1002/zaac.200700535.

24. Biehl, E.; Deiseroth, H.J. Darstellung, Strukturchemie und Magnetismus der Amalgame MHg$_{11}$ (M: K, Rb, Ba, Sr). *Z. Anorg. Allg. Chem.* **1999**, *625*, 1073–1080, doi:10.1002/(SICI)1521-3749(199907)625:7<1073::AID-ZAAC1073>3.0.CO;2-V.

25. Hoch, C.; Simon, A. Tetramethylammoniumamalgam [N(CH$_3$)$_4$]Hg$_8$. *Z. Anorg. Allg. Chem.* **2006**, *632*, 2288–2294, doi:10.1002/zaac.200600163.

26. Tambornino, F.; Hoch, C. Bad metal behaviour in the new Hg-rich amalgam KHg$_6$ with polar metallic bonding. *J. Alloys Compd.* **2015**, *818*, 299–304, doi:10.1016/j.jallcom.2014.08.173.

27. Lahiri, S.K.; Angilello, J.; Natan, M. Precise lattice parameter determination of PtHg$_4$. *J. Appl. Crystallogr.* **1982**, *15*, 100–101, doi:10.1107/S0021889882011443.

28. Biehl, E.; Deiseroth, H.J. K$_2$Hg$_7$ und Rb$_2$Hg$_7$, zwei Vertreter eines neuen Strukturtyps binärer intermetallischer Verbindungen. *Z. Anorg. Allg. Chem.* **1999**, *625*, 1337–1342, doi:10.1002/(SICI)1521-3749(199908)625:8<1337::AID-ZAAC1337>3.0.CO;2-W.

29. Tambornino, F. Electrolytic Synthesis and Structural Chemistry of Intermetallic Phases with Polar Metal-Metal Bonding. Ph.D. Thesis, LMU, München, Germany, 2016.

30. Merlo, F.; Fornasini, M.L. Crystal structure of the R$_{11}$Hg$_{45}$ compounds (R = La, Ce, Pr, Nd, Sm, Gd, U). *J. Less-Common Met.* **1976**, *44*, 259–265, doi:10.1016/0022-5088(79)90173-5.

31. Iandelli, A.; Palenzona, A. Su alcuni composti intermetallici dell'europio con zinco, cadmio e mercurio. *Atti Accad. Nazl. Lincei Rend. Cl. Sci. Fis. Mater. Nat.* **1964**, *37*, 165–168.

32. Hoch, C.; Simon, A. Na$_{11}$Hg$_{52}$: Komplexität in einem polaren Metall. *Angew. Chem.* **2012**, *124*, 3316–3319, doi:10.1002/ange.201108064.

33. Hoch, C.; Simon, A. Na$_{11}$Hg$_{52}$: Complexity in a polar metal. *Angew. Chem. Int. Ed.* **2012**, *51*, 3262–3265, doi:10.1002/anie.201108064.

34. Stoe & Cie. *X-SHAPE v. 2.07*; Stoe & Cie: Darmstadt, Germany, 2005.

35. Stoe & Cie. *X-RED v. 1.31*; Stoe & Cie: Darmstadt, Germany, 2005.

36. Sheldrick, G.M. A short history of SHELX. *Acta Crystallogr.* **2008**, *A64*, 112–122, doi:10.1107/S0108767307043930.

37. Gelato, L.M.; Parthé, E. Structure Tidy—A computer program to standardize crystal structure data. *J. Appl. Crystallogr.* **1987**, *20*, 139–143, doi:10.1107/S0021889887086965.

38. Rahlfs, P. Die Kristallstruktur des Ni$_3$Sn (Mg$_3$Cd-Typ = Überstruktur der hexagonal dichtesten Kugelpackung). *Metallwirtschaft* **1937**, *16*, 343–345.

39. Lyubimtsev, A.L.; Baranov, A.I.; Fischer, A.; Kloo, L.; Popovkin, B.A. The structures and bonding of Ni$_3$Sn. *J. Alloys Compd.* **2002**, *340*, 167–172, doi:10.1016/S0925-8388(02)00047-6.

40. Lihl, F.; Krinbauer, H. Untersuchung binärer metallischer Systeme mit Hilfe des Amalgamverfahrens. Das System Nickel-Zinn. *Monatsh. Chem.* **1955**, *86*, 745–751, doi:10.1007/BF00902566.

41. Watanabe, Y.; Murakumi, Y.; Kachi, S. Martensitic and massive transformations and phase diagram in Ni$_{3-x}$M$_x$Sn (M = Cu, Mn) alloys. *J. Jpn. Inst. Met.* **1981**, *45*, 551–558, doi:10.2320/jinstmet1952.45.6_551.

42. Mantina, M.; Chamberlin, A.C.; Valero, R.; Cramer, C.J.; Truhlar, D.G. Consistent van der Waals radii for the whole main group. *J. Phys. Chem.* **2009**, *113*, 5806–5812, doi:10.1021/jp8111556.

43. Köhler, J.; Whangbo, M.-H. Late transition metal anions acting as p-metal elements. *Solid State Sci.* **2008**, *10*, 444–449, doi:10.1016/j.solidstatesciences.2007.12.001.

44. Barrett, C.S. The structure of mercury at low temperature. *Acta Crystallogr.* **1957**, *10*, 58–60, doi:10.1107/S0365110X57000134.

45. Deng, S.; Simon, A.; Köhler, J. Supraleitung und chemische Bindung in Quecksilber. *Angew. Chem.* **1998**, *110*, 664–666, doi:10.1002/(SICI)1521-3757(19980302)110:5<664::AID-ANGE664>3.0.CO;2-8.

46. Deng, S.; Simon, A.; Köhler, J. Superconductivity and chemical bonding in mercury. *Angew. Chem. Int. Ed.* **1998**, *37*, 640–643, doi:10.1002/(SICI)1521-3773(19980316)37:5<640::AID-ANIE640>3.0.CO;2-G.

47. Zintl, E.; Schneider, A. Röntgenanalyse der Lithium-Amalgame. *Z. Elektrochem. Angew. Phys. Chem.* **1935**, *41*, 771–774, doi:10.1002/bbpc.19350411105.

48. Bruzzone, G.; Merlo, F. The strontium-mercury system. *J. Less-Common Met.* **1974**, *35*, 153–157, doi:10.1016/0022-5088(74)90154-4.

49. Bruzzone, G.; Ruggiero, A.F. Struttura di alcuni composti intermetallici dell'ittrio—I. Composti con Cu, Ag, Au, Zn, Cd, Hg. *Atti Accad. Nazl. Lincei Rend. Cl. Sci. Fis. Mater. Nat.* **1962**, *33*, 312–314.

50. Laube, E.; Nowotny, H.N. Die Kristallstrukturen von ScHg, ScHg$_3$, YCd, YHg und YHg$_3$. *Monatsh. Chem.* **1963**, *94*, 851–858, doi:10.1007/BF00902359.

51. Bruzzone, G.; Merlo, F. The lanthanum-mercury system. *J. Less-Common Met.* **1976**, *44*, 259–265, doi:10.1016/0022-5088(76)90140-5.

52. Olcese, G.L. Sul comportamento magnetico del cerio nei composti intermetallici. II. I sistemi Ce-Zn, Ce-Cd Ce-Hg. *Atti Accad. Nazl. Lincei Rend. Cl. Sci. Fis. Mater. Nat.* **1963**, *35*, 48–52.

53. Palenzona, A. MX$_3$ intermetallic phase of the rare earths with Hg, In, Tl, Pb. *J. Less-Common Met.* **1966**, *10*, 290–292, doi:10.1016/0022-5088(66)90031-2.

54. Ferro, R. The crystal structures of ThHg$_3$, ThIn$_3$, ThTl$_3$, ThSn$_3$ and ThPb$_3$. *Acta Crystallogr.* **1958**, *11*, 737–738, doi:10.1107/S0365110X5800195X.

55. Rundle, R.E.; Wilson, A.S. The structures of some metal compounds of uranium. *Acta Crystallogr.* **1949**, *2*, 148–150, doi:10.1107/S0365110X49000400.

56. Tambornino, F.; Hoch, C. The Hg-richest europium amalgam, Eu$_{10}$Hg$_{55}$. *Z. Anorg. Allg. Chem.* **2015**, *641*, 537–542, doi:10.1002/zaac.201400561.

57. Tambornino, F.; Sappl, J.; Pultar, F.; Cong, T.M.; Hübner, S.; Giftthaler, T.; Hoch, C. Electrocrystallization—A synthetic method for intermetallic phases with polar metal-metal bonding. *Inorg. Chem.* **2016**, *55*, 11551–11559, doi:10.1021/acs.inorgchem.6b02068.

58. Stoe & Cie. *X-AREA v. 1.39*; Stoe & Cie: Darmstadt, Germany, 2006.

crystals

MDPI

Article

Synthesis, Crystal Structure and Luminescent Properties of 2D Zinc Coordination Polymers Based on Bis(1,2,4-triazol-1-yl)methane and 1,3-Bis(1,2,4-triazol-1-yl)propane

Evgeny Semitut [1,2], Taisiya Sukhikh [2,3], Evgeny Filatov [2,3], Alexey Ryadun [2] and Andrei Potapov [1,*]

[1]   Department of Biotechnology and Organic Chemistry, National Research Tomsk Polytechnic University, 30 Lenin Ave., 634050 Tomsk, Russia; semitut@niic.nsc.ru
[2]   Nikolaev Institute of Inorganic Chemistry, Siberian Branch of the Russian Academy of Sciences, Lavrentieva Ave. 3, 630090 Novosibirsk, Russia; sukhikh@niic.nsc.ru (T.S.); decan@niic.nsc.ru (E.F.); ryadunalexey@mail.ru (A.R.)
[3]   Department of Natural Sciences, Novosibirsk State University, Pirogova Str. 2, 630090 Novosibirsk, Russia
*   Correspondence: potapov@tpu.ru; Tel.: +7-923-403-4103

Academic Editor: Matthias Weil
Received: 8 November 2017; Accepted: 27 November 2017; Published: 29 November 2017

Abstract: Two new two-dimensional zinc(II) coordination polymers containing 2,5-thiophenedicarboxylate and bitopic ligands bis(1,2,4-triazol-1-yl)methane (btrm) or 1,3-bis(1,2,4-triazol-1-yl)propane (btrp) were synthesized. Synthesized compounds were characterized by IR spectroscopy, elemental analysis, powder X-ray diffraction, and thermal analysis. Crystal structures of coordination polymers were determined and their structural peculiarities are discussed. The differences in structural features, thermal behavior, and luminescent properties are discussed.

Keywords: zinc; coordination polymer; bitopic ligand; crystal structure; thermal analysis; luminescence; 2,5-thiophenedicarboxylic acid; bis(1,2,4-triazol-1-yl)methane; 1,3-bis(1,2,4-triazol-1-yl)propane

1. Introduction

Coordination polymers and metal-organic frameworks attract the unceasing attention of researchers due to their wide range of potential applications [1–6]. One of the universal approaches to the construction of metal-organic frameworks uses a combination of metal ions with aromatic di- or polycarboxylate donors and rigid or flexible N-donor bitopic ligands [7,8]. Among dicarboxylic acids, 2,5-thiophenedicarboxylic acid (H_2tdc, Figure 1) was used to enhance the gas sorption properties of MOFs [9–11], fine-tune the topology of the constructed coordination polymers [12–14], prepare coordination polymers with luminescent properties [9,10,15] including those suitable for LED [16] applications, and sense metal ions and small molecules [17,18]. Metal-organic frameworks based on tdc^{2-} donors demonstrating potential photocatalytic [19,20] and magnetic [21] applications were also reported. Bitopic heterocyclic ligands based on semi-rigid di(1,2,4-triazol-1-yl) derivatives [22] or flexible bis(imidazol-1-yl)alkanes [19,23–28] are usually used in combination with tdc^{2-} and metal ions to build coordination networks. Despite a very large number of reported bis(imidazol-1-yl)alkane-linked frameworks based on H_2tdc, no examples of coordination polymers with structurally similar bis(1,2,4-triazol-1-yl)alkanes have been prepared so far. There are a number of publications reporting the study of zinc coordination polymers based on 1,3-bis(1,2,4-triazol-1-yl)propane (btrp, Figure 1) and aromatic di-, tri-, and tetracarboxylates [29–43], and several examples of 1D coordination polymers [44,45] and discrete complexes [44,46,47] have also been reported. Zinc coordination chemistry with bis(1,2,4-triazol-1-yl)methane (btrm, Figure 1) is much less studied, and only two examples of 1D coordination polymers have been reported so far [48,49].

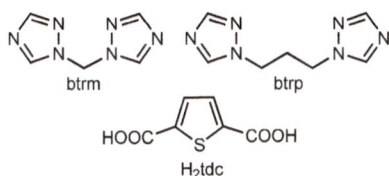

Figure 1. Bis(1,2,4-triazol-1-yl)methane, 1,3-bis(1,2,4-triazol-1-yl)propane and 2,5-thiophenedicarboxylic acid ligands used in this work for the preparation of coordination polymers.

In order to explore the possibility of the preparation of new coordination polymers based on bis(1,2,4-triazol-1-yl) and H_2tdc ligands with enhanced functional properties, we have studied the reaction between zinc nitrate, H_2tdc, and btrm or btrp linkers. As a result, the first examples of bis(1,2,4-triazol-1-yl)methane and 1,3-bis(1,2,4-triazol-1-yl)propane-linked zinc -2,5-thiophenedicarboxylate coordination polymers were prepared, and their crystal structures, thermal behavior, and luminescent properties were investigated.

2. Results and Discussion

2.1. Synthesis of Coordination Polymers

The coordination polymers **1** and **2** were characterized by thermal analysis, single crystal and powder X-ray diffraction methods, CHNS analysis, and IR spectroscopy. In addition, their photoluminescence properties were investigated.

Syntheses of coordination polymers by the reaction of zinc nitrate, btrp or btrm ligands, and 2,5-thiophenedicarboxylic acid (H_2tdc) were carried out under solvothermal conditions at 95 °C in dimethylformamide (DMF). Zn-ligand-H_2tdc ratios remained constant and equimolar in all experiments. The duration of heating was varied from 12 to 36 h in order to optimize the yield and purity of the crystalline product.

The reaction of equimolar amounts of zinc nitrate, btrm, and H_2tdc in the DMF solution (Zn^{2+} concentration 1.0 M) at 95 °C for 24 h gave prismatic crystals of coordination polymer [Zn(btrm)(tdc)]·nDMF **1**. The powder XRD analysis has shown that carrying out the reaction for a longer period of time (e.g., for 36 h) results in the formation of the additional unidentified product as an impurity. This impurity can be removed by washing the precipitate with warm DMF. The XRD patterns for compound **1** and the additional by-product are shown in Figure S1.

When equimolar amounts of zinc nitrate, btrp and H_2tdc were heated in DMF solution (Zn^{2+} concentration 1.0 M) at 95 °C for 12 h, and coordination polymer [Zn(btrp)(tdc)]·nDMF **2** as prismatic crystals suitable for X-ray structure determination was obtained. The powder XRD analyses of the polycrystalline sample in comparison with those simulated from single crystal data patterns are shown in Figure S2. The IR spectra of both compounds contain characteristic bands associated with vibrations of bis(triazol-1-yl) ligands and coordinated 2,5-thiophenedicarboxylate anions (Figure S3).

2.2. Crystal Structures

2.2.1. Crystal Structure of Polymer [Zn(tdc)(btrm)]·nDMF (**1**)

The complex [Zn(tdc)(btrm)]·nDMF (**1**) is a 2D coordination polymer. The Zn atom coordinates two crystallographically independent (tdc)$^{2-}$ ligands (halves) to form chains (Figure 2a). Analysis of bond lengths reveals the shortest distances for Zn1–O11 of 1.94 and Zn1–O21 of 1.99 Å typical for this type of coordination compound. Both (tdc)$^{2-}$ anions act as (μ-O)$_2$-coordinating ligands. Zn1$\cdots$O12 (2.99 Å) and Zn1$\cdots$O22 (2.65 Å) distances are much longer, although the latter can be considered as a long range interaction in agreement with the analysis of normalized contact distances (d_{norm}) on a Hirshfeld surface [50,51] (Figure S4a). In contrast to atom O22, atom O12 reveals contact with the H of the neighboring triazole unit, with a corresponding distance of 2.25 Å (Figure S4b). {Zn(tdc)} chains are linked by (μ-N)$_2$-coordinating btrm ligands to form corrugated layers arranged along the *ac* plane (Figure 2b). Within the layer, Zn atoms deviate from their mean plane significantly by 3.54 Å (Figure S6). These layers are stacked one above the other (Figure 2c) leaving channel voids of *ca.* 35% filled by highly disordered DMF solvent molecules (Figure 3).

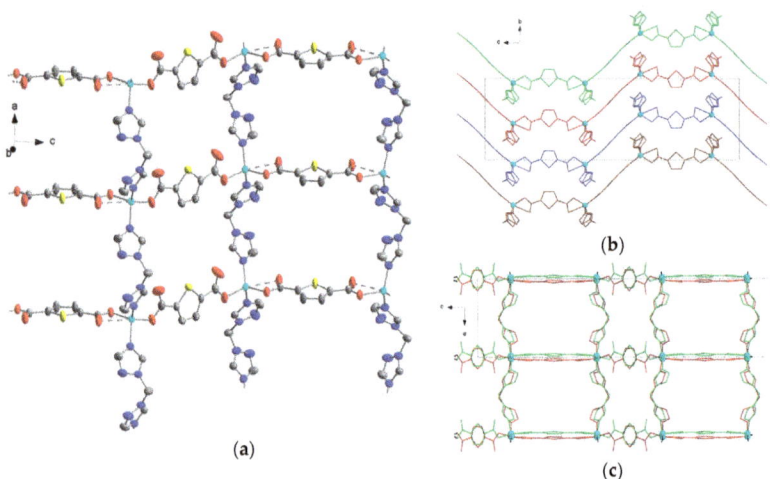

(a)

(b)

(c)

Figure 2. (a) Displacement ellipsoid plot of complex [Zn(tdc)(btrm)]·nDMF (**1**) showing 50% probability ellipsoids. H atoms are not shown for clarity. Dashed lines indicate long-range Zn$\cdots$O interactions; (b,c) Relative arrangement of the layers of **1** colored green, red, blue, and brown.

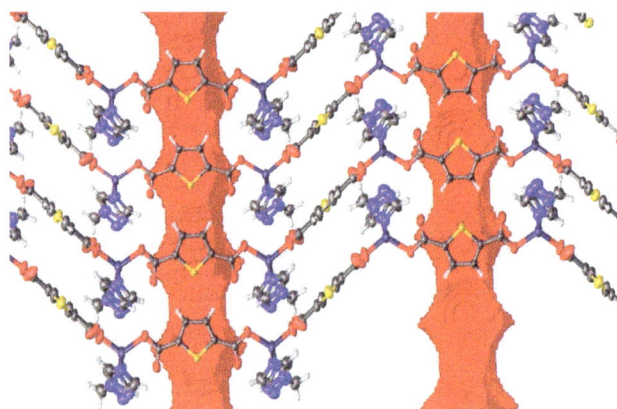

Figure 3. Representation of channel voids in the structure of the complex [Zn(tdc)(btrm)]·nDMF (**1**).

2.2.2. Crystal Structure of Polymer [Zn(tdc)(btrp)]·nDMF (2)

The complex [Zn(tdc)(btrp)]·nDMF (**2**) also has a 2D structure (Figure 4a). Similar to **1**, a four-connected net of Zn atoms is observed, but (tdc)$^{2-}$ and btrp ligands alternate along the chains. The whole layer is arranged along the *ab* plane and it less corrugated compared to layers of polymer **1**. Within the layer, Zn atoms deviate from their mean plane by 0.55 Å (Figure S7). Similar to compound **1**, the dicarboxylate ligand in complex **2** adopts a (μ-O)$_2$-coordination mode, revealing pairs of short (1.93, 1.98 Å) and long (2.61, 3.29 Å) Zn⋯O distances. According to the analysis of the d_{norm} map on the Hirshfeld surface, the O14 atom shows weak interaction with the Zn, as well as with the H, atom of the triazole unit with corresponding Zn⋯O and H⋯O distances of 2.61 and 2.53 Å (Figure S5). On the contrary, the O12 atom reveals no close contacts with Zn and H atoms. Crystal packing differences are observed for compounds **1** and **2**. Coordination polymer **2** shows a double interpenetration of the layers with each Zn node of one array lying above or below the approximate center of the space of another layer (Figure 4b,c). Due to interpenetration, only separate voids filled by DMF molecules are revealed (Figure 5), with the solvent accessible volume (of *ca.* 22%) being lower than in **1**. Each void contains one DMF molecule, which is disordered by two positions due to its proximity to an inversion center.

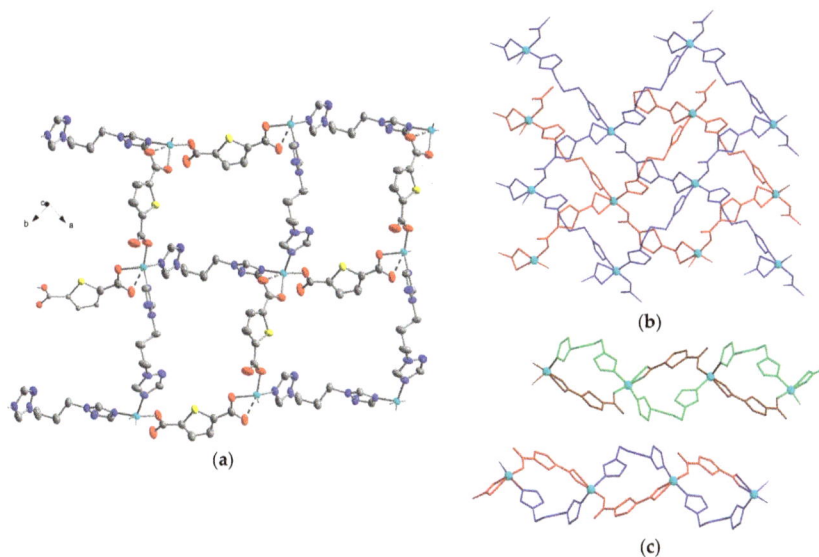

Figure 4. (a) Displacement ellipsoid plot of complex [Zn(tdc)(btrp)]·nDMF (**2**) showing 50% probability ellipsoids. H atoms are not shown for clarity. Dashed lines indicate long-range Zn⋯O interactions; (b,c). Relative arrangement of the layers of **1** colored green, red, blue, and brown.

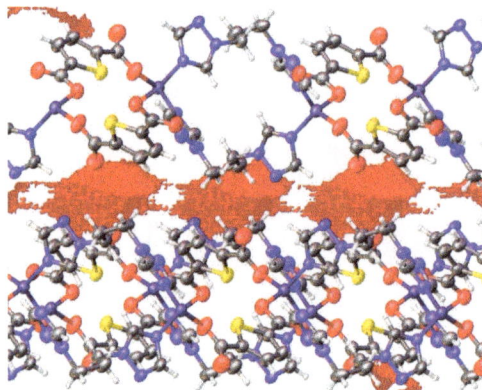

Figure 5. Representation of the voids in the structure of the complex [Zn(tdc)(btrp)]·nDMF (**2**).

2.3. Thermal Analysis

The analysis of the thermal properties of synthesized compounds revealed that the processes of removing guest molecules from coordination polymers **1** and **2** have some significant differences (Figure 6a,b). The process of desolvation for compound **1** starts at about 120 °C, while the desolvation process for **2** starts almost at room temperature. The number of guest molecules for **2** is variable and decreases under storage in air revealed by the results of CHNS analysis of the samples stored in air for a few weeks. The second step for both compounds starts at about 250 °C and lasts up to about 400 °C for **1** and up to 500 °C for **2**.

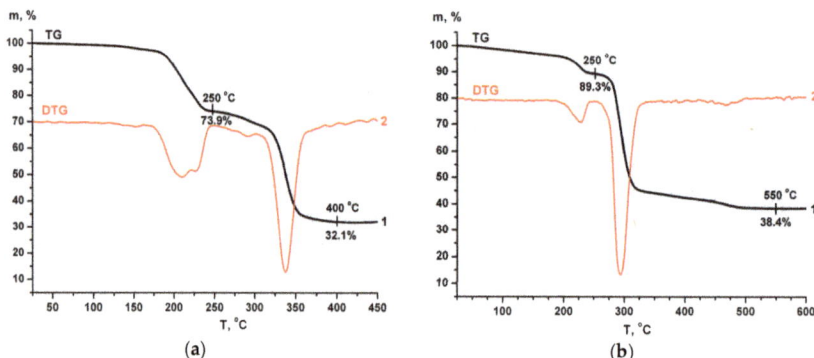

Figure 6. Curves of thermal analysis for compound **1** (**a**) and for compound **2** (**b**).

2.4. Luminescent Properties

The luminescence and excitation spectra of compounds **1** and **2** are shown in Figure 7. Upon excitation at 375 nm (for **1**) and at 330 nm (for **2**), the photoluminescence spectra demonstrate wide bands with maxima at 430 and 440 nm, respectively. The band of **1** is vibrationally resolved with two shoulders having bathochromic and hypsochromic shifts. The excitation spectrum of **1** also has a vibrational resolution. The ligands btrm and btrp (for images of btrp spectra, see ref. [43]) reveal broad emission bands with the maxima at 410 and 440 nm, respectively. The excitation bands of coordination polymers are relatively narrow, while for the ligands, these bands are wide. It is interesting to note that the quantum yield (QY) for **1** is several times higher than the QY of the btrm ligand. On the contrary, the QY for **2** decreases by a few orders compared to the btrp ligand (Table 1).

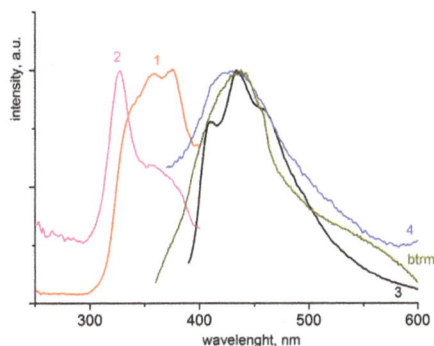

Figure 7. Normalized emission (λ_{ex} = 330 nm and 375 nm) and excitation spectra: 1, 3—compound **1**; 2, 4—compound **2**. Normalized emission (λ_{ex} = 330 nm) spectrum of btrm.

Table 1. Photoluminescence data for coordination polymers **1, 2** and ligands btrm, btrp.

	1	2	Btrm	Btrp
Ex, nm	330	375	330	260 br
Em, nm	410 (sh), 430, 460 (sh)	440	440	410
QY	0.1	<0.005 [1]	0.03	0.04

[1] QY is too low to be measured.

3. Experimental Section

3.1. Materials and Methods

The starting reagents used for the synthesis of coordination compounds—$Zn(NO_3)_2 \cdot 6H_2O$ (chemical grade) and dimethyl formamide (analytical grade)—were used as received. Btrm and btrp ligands were prepared as reported previously [43]. Elemental analysis was carried out on a Eurovector EuroEA 3000 analyzer (Eurovector SPA, Redavalle, Italy). Infrared (IR) spectra of solid samples as KBr pellets were recorded on a FT-801 spectrometer (550–4000 cm^{-1}, Kailas OU, Tallin, Estonia). Polycrystalline samples were studied in 2θ range 5°–60° on a DRON RM4 powder diffractometer (Burevestnik, Saint Petersburg, Russia) equipped with a CuKα source (λ = 1.5418 Å) and graphite monochromator for the diffracted beam. Indexing of the diffraction patterns was done using data for compounds reported in the JCPDS-ICDD database [52].

The thermal stability of coordination polymers was studied in inert (He) atmosphere. Thermogravimetric measurements were carried out on a NETZSCH thermobalance TG 209 F1 Iris (Erich NETZSCH GmbH & Co. Holding KG, Selb, Germany). Open Al_2O_3 crucibles were used (loads 5–10 mg, heating rate 10 $K \cdot min^{-1}$).

Room temperature excitation and emission spectra were recorded with a Horiba Jobin Yvon Fluorolog 3 photoluminescence spectrometer (Horiba Jobin Yvon, Edison, NJ, USA) equipped with a 450 W ozone-free Xe-lamp, cooled PC177CE-010 photon detection module with a PMT R2658, and double grating excitation and emission monochromators. Powdered samples for measurements were placed between two non-fluorescent quartz plates. Quantum yields were determined using a Quanta-φ integrating sphere. Excitation and emission spectra were corrected for source intensity (lamp and grating) and emission spectral response (detector and grating) by standard correction curves.

3.2. X-ray Structure Determination

Single-crystal XRD data for the complexes **1** and **2** were collected on a Bruker Apex DUO diffractometer (Bruker Corporation, Billerica, MA, USA) equipped with a 4K CCD area detector at 298(2) K using graphite-monochromated MoKα radiation (λ = 0.71073 Å) (Table 2). The φ- and ω-scan techniques were employed to measure intensities. Absorption corrections were applied with the use of the SADABS program [53]. The crystal structures were solved by direct methods and refined by full-matrix least squares techniques with the use of the SHELXTL package [54]. Atomic thermal displacement parameters for non-hydrogen atoms were refined anisotropically. The positions of hydrogen atoms were calculated corresponding to their geometrical conditions and refined using the riding model. DFIX, DANG restrains, and EADP constrains were applied to atoms of the disordered DMF molecules of **2**. In compound **1**, solvent molecules displayed unresolvable disorder. Therefore, the structure was treated *via* the PLATON/SQUEEZE [55] procedure to remove the contribution of the electron density in the solvent regions from the intensity data. The total potential solvent accessible void volume was estimated to be *ca.* 1300 Å^3 and the electron count per unit cell was 380, which were assigned to eight DMF molecules per unit cell and one molecule per formula unit.

The Hirshfeld promolecular surface mapped over d_{norm} plots of the complexes **1** and **2** was built using the Crystal Explorer (version 17.5) program [56].

Table 2. Crystallographic data for compounds **1** and **2**.

Parameter	Compound 1	Compound 2
Empirical formula	$C_{11}H_8N_6O_4SZn$	$(C_{13}H_{12}N_6O_4SZn)\cdot0.5(C_3H_7NO)$
Formula weight	385.66	450.26
Crystal system	orthorhombic	monoclinic
Space group	*Pbcm*	$P2_1/c$
Unit cell dimensions *a*, Å	9.3189(3)	6.4235(2)
b, Å	11.5387(3)	15.4396(5)
c, Å	37.2569(11)	19.2190(7)
β, °		90.7550(10)
Volume, Å^3	4006.2(2)	1905.90(11)
Z	8	4
Density (calcd.), g·cm^{-3}	1.279	1.569
F(000)	1552	920
Abs. coefficient, mm^{-1}	1.352	1.436
Crystal size, mm^3	0.18 × 0.17 × 0.08	0.40 × 0.30 × 0.10
$2\theta_{max}$, °	51.60	51.60
Index range	$-11 \le h \le 7$ $-14 \le k \le 14$ $-45 \le l \le 45$	$-7 \le h \le 7$ $-18 \le k \le 18$ $-23 \le l \le 23$
Reflections collected	42,882	29,130
Independent reflections	3900 [*R*(int) = 0.0416]	3405 [*R*(int) = 0.0381]
Completeness to 2θ = 50.5, %	99.8	99.4
Reflections, $I \ge 2\sigma(I)$	3201	3405
Parameters	210	244
Final R indices [$I > 2\sigma(I)$]	R1 = 0.0356 wR2 = 0.0836	R1 = 0.0356 wR2 = 0.1064
R indices (all data)	R1 = 0.0470 wR2 = 0.0871	R1 = 0.0378 wR2 = 0.1084
GoF	1.048	1.060
Residual electron density (min/max, e/Å^3)	−0.282/0.310	−0.613/0.952

Crystallographic data for the structural analysis have been deposited with the Cambridge Crystallographic Data Centre, CCDC Nos. 1584008 for compound **1** and 1584007 for compound **2**. Copies of the data can be obtained free of charge from the Cambridge Crystallographic Data Centre, 12 Union Road, Cambridge CB2 1EZ, UK (Fax: +44-1223-336-033; e-mail: deposit@ccdc.cam.ac.uk).

3.3. Synthesis of Compounds

3.3.1. Synthesis of [Zn(btrm)(tdc)]·nDMF (**1**)

Solution of $Zn(NO_3)_2 \cdot 6H_2O$ (1.0 M, 0.64 mL, 0.64 mmol) in DMF was added to the mixture of btrm ligand (96 mg, 0.64 mmol) and 1.6 mL of 0.4 M H_2tdc (0.64 mmol) solution in DMF in a glass vial. The mixture was stirred for several minutes at room temperature until the complete dissolution of all reagents. The vial was placed in an oven at 95 °C for 24 h. Then, the vial was removed from the oven and cooled to room temperature. Colorless prismatic crystals were formed on the bottom. The crystals were washed twice with 2 mL of DMF and then stored in the glass vial under DMF. The yield was about 120 mg (*ca.* 40%). IR bands, cm^{-1}: 3124, 2930, 1671, 1610, 1529, 1469, 1374, 1289, 1210, 1120, 1032, 990, 909, 859, 819, 780, 745, 673, 638. Elemental analysis: found, %: C 36.1, H 3.4, N 21.2, S 7.2; calculated ([Zn(btrm)(tdc)]·nDMF, n = 1), %: C 36.7, H 3.3, N 21.4, S 7.0.

3.3.2. Synthesis of [Zn(btrp)(tdc)]·nDMF (**2**)

Solution of $Zn(NO_3)_2 \cdot 6H_2O$ (1.0 M, 0.5 mL, 0.5 mmol) in DMF was added to the mixture of btrm ligand (89.1 mg, 0.5 mmol) and 1.25 mL of 0.4 M H_2tdc (0.5 mmol) solution in DMF in a glass vial. The mixture was stirred for several minutes at room temperature until the complete dissolution of all reagents. The vial was placed in an oven at 95 °C for 12 h. Then, the vial was removed from the oven and cooled to room temperature. Colorless prismatic crystals formed on the bottom. The crystals were washed twice with 2 mL of DMF and then stored in the glass vial under DMF. The yield was about 100 mg (*ca.* 40%). IR bands, cm^{-1}: 3125, 2934, 1668, 1630, 1590, 1531, 1461, 1353, 1280, 1209, 1173, 1128, 995, 899, 812, 769, 677, 651, 587. Elemental analysis: found, %: C 39.1, H 3.5, N 20.0, S 6.6; calculated ([Zn(btrp)(tdc)]·nDMF, n = 1), %: C 39.5, H 3.9, N 20.1, S 6.6.

4. Conclusions

In summary, this study presents the two first examples of zinc coordination polymers with 2,5-thiophenedicarboxylic acid, in which bis(1,2,4-triazol-yl)methane or 1,3-bis(1,2,4-triazol-yl)propane play the role of auxiliary ligands. Both compounds are two-dimensional coordination polymers. In the case of a shorter btrm linker, the layers are stacked above each other forming channels (*ca.* 35 % of cell volume) filled with DMF molecules. The use of longer and more flexible btrp linkers results in the formation of doubly-interpenetrated layers with closed voids (*ca.* 22 % of cell volume). Thermal analysis of coordination polymers has shown that the solvate molecules are easier to remove in the case of the btrm linker. The quantum yield of **1** was found to be several times higher than that of the btrm ligand. On the contrary, the quantum yield of **2** decreases by a few orders relative to the btrp ligand.

Supplementary Materials: The following are available online at www.mdpi.com/2073-4352/7/12/354/s1, Figure S1: XRD patterns of compound **1**, Figure S2: XRD patterns of compound **2**, Figure S3: IR spectra of coordination polymers **1** and **2**, Figure S4: (**a,b**) The d_{norm} Hirshfeld surface of $(tdc)^{2-}$ ligands of the complex [Zn(tdc)(btrm)]·0.5DMF (**1**), Figure S5: The d_{norm} Hirshfeld surface of $(tdc)^{2-}$ ligand of the complex [Zn(tdc)(btrp)]·0.5DMF (**2**) in different projections, Figure S6: Fragment of the layer of the complex **1**, Figure S7: Fragment of the layer of the complex **2**.

Acknowledgments: The reported study was supported by the Russian Science Foundation, grant No. 15-13-10023, and the analytical characterization of 1,2,4-triazole derivatives was carried out with support from Tomsk Polytechnic University Competitiveness Enhancement Program grant, project number TPU CEP_IHTP_73\2017.

Crystals **2017**, *7*, 354

Author Contributions: Evgeny Semitut and Andrei Potapov conceived and designed the experiments; Evgeny Semitut and Taisiya Sukhikh carried out the synthesis; Taisiya Sukhikh performed X-ray structure determination and analyzed the results; Evgeny Filatov performed X-Ray powder diffraction analysis; and Alexey Ryadun investigated luminescent properties. All authors took part in writing and discussion processes.

Conflicts of Interest: The authors declare no conflict of interest.

References

1. Pettinari, C.; Marchetti, F.; Mosca, N.; Tosi, G.; Drozdov, A. Application of metal-organic frameworks. *Polym. Int.* **2017**, *66*, 731–744. [CrossRef]
2. You, A.; Li, Y.; Zhang, Z.M.; Zou, X.Z.; Gu, J.Z.; Kirillov, A.M.; Chen, J.W.; Chen, Y.B. Novel metal-organic and supramolecular 3D frameworks constructed from flexible biphenyl-2,5,3'-tricarboxylate blocks: Synthesis, structural features and properties. *J. Mol. Struct.* **2017**, *1145*, 339–346. [CrossRef]
3. Gu, J.-Z.; Cai, Y.; Liu, Y.; Liang, X.-X.; Kirillov, A.M. New lanthanide 2D coordination polymers constructed from a flexible ether-bridged tricarboxylate block: Synthesis, structures and luminescence sensing. *Inorg. Chim. Acta* **2018**, *469*, 98–104. [CrossRef]
4. Huang, W.; Pan, F.; Liu, Y.; Huang, S.; Li, Y.; Yong, J.; Li, Y.; Kirillov, A.M.; Wu, D. An efficient blue-emissive metal-organic framework (MOF) for lanthanide-encapsulated multicolor and stimuli-responsive luminescence. *Inorg. Chem.* **2017**, *56*, 6362–6370. [CrossRef] [PubMed]
5. Semitut, E.; Komarov, V.; Sukhikh, T.; Filatov, E.; Potapov, A. Synthesis, crystal structure and thermal stability of 1D linear silver(I) coordination polymers with 1,1,2,2-Tetra(pyrazol-1-yl)ethane. *Crystals* **2016**, *6*, 138. [CrossRef]
6. Barsukova, M.; Goncharova, T.; Samsonenko, D.; Dybtsev, D.; Potapov, A. Synthesis, crystal structure, and luminescent properties of new zinc(II) and cadmium(II) metal-organic frameworks based on flexible bis(imidazol-1-yl)alkane ligands. *Crystals* **2016**, *6*, 132. [CrossRef]
7. Batten, S.R.; Champness, N.R. Coordination polymers and metal-organic frameworks: Materials by design. *Philos. Trans. R. Soc. A Math. Phys. Eng. Sci.* **2016**, *375*. [CrossRef] [PubMed]
8. Dybtsev, D.N.; Sapianik, A.A.; Fedin, V.P. Pre-synthesized secondary building units in the rational synthesis of porous coordination polymers. *Mendeleev Commun.* **2017**, *27*, 321–331. [CrossRef]
9. An, Z.; Wang, J. Gas adsorption and luminescent properties of a porous pcu-type Zn(II) coordination polymer. *Synth. React. Inorg. Met. Nano-Met. Chem.* **2016**, *46*, 1810–1814. [CrossRef]
10. An, Z.; Zhu, L. A new luminescent Zn(II) coordination polymer with eightfold interpenetrated ths topology. *Synth. React. Inorg. Met. Nano-Metal Chem.* **2016**, *46*, 1367–1370. [CrossRef]
11. Lu, Y.; Dong, Y.; Qin, J. Porous pcu-type Zn(II) framework material with high adsorption selectivity for CO_2 over N_2. *J. Mol. Struct.* **2016**, *1107*, 66–69. [CrossRef]
12. Erer, H. Effect of O and S heteroatom containing heterocyclic dicarboxylates in the structural diversity of cadmium(II) coordination polymers with flexible 1-substituted (1,2,4-triazole) ligand. *Polyhedron* **2015**, *102*, 201–206. [CrossRef]
13. Li, X.; Zhou, P.; Dong, Y.; Liu, H. Structural diversity of a series of 2D Zn(II) coordination polymers tuned by different dicarboxylic acids ligands. *J. Inorg. Organomet. Polym. Mater.* **2015**, *25*, 650–656. [CrossRef]
14. Sapchenko, S.A.; Samsonenko, D.G.; Fedin, V.P. Synthesis, structure and luminescent properties of metal-organic frameworks constructed from unique Zn- and Cd-containing secondary building blocks. *Polyhedron* **2013**, *55*, 179–183. [CrossRef]
15. Sapchenko, S.A.; Saparbaev, E.S.; Samsonenko, D.G.; Dybtsev, D.N.; Fedin, V.P. Synthesis, structure, and properties of a new layered coordination polymer based on Zinc(II) carboxylate. *Russ. J. Coord. Chem.* **2013**, *39*, 549–552. [CrossRef]
16. Xie, W.; He, W.-W.; Du, D.-Y.; Li, S.-L.; Qin, J.-S.; Su, Z.-M.; Sun, C.-Y.; Lan, Y.-Q. A stable Alq3@MOF composite for white-light emission. *Chem. Commun.* **2016**, *52*, 3288–3291. [CrossRef] [PubMed]
17. Gu, T.-Y.; Dai, M.; Young, D.J.; Ren, Z.-G.; Lang, J.-P. Luminescent Zn(II) coordination polymers for highly selective sensing of Cr(III) and Cr(VI) in water. *Inorg. Chem.* **2017**, *56*, 4668–4678. [CrossRef] [PubMed]

18. Liu, F.-H.; Qin, C.; Ding, Y.; Wu, H.; Shao, K.-Z.; Su, Z.-M. Pillared metal organic frameworks for the luminescence sensing of small molecules and metal ions in aqueous solutions. *Dalton Trans.* **2015**, *44*, 1754–1760. [CrossRef] [PubMed]

19. Zhang, C.; Ma, D.; Zhang, X.; Ma, J.; Liu, L.; Xu, X. Preparation, structure and photocatalysis of metal-organic frameworks derived from aromatic carboxylate and imidazole-based ligands. *J. Coord. Chem.* **2016**, *69*, 985–995. [CrossRef]

20. Zhao, S.; Li, M.; Shi, L.-L.; Li, K.; Li, B.-L.; Wu, B. Syntheses, structures and photocatalytic properties of three copper(II) coordination polymers. *Inorg. Chem. Commun.* **2016**, *70*, 185–188. [CrossRef]

21. Zhou, L.; Wang, C.; Zheng, X.; Tian, Z.; Wen, L.; Qu, H.; Li, D. New metal-organic frameworks based on 2,5-thiophenedicarboxylate and pyridine- or imidazole-based spacers: Syntheses, topological structures, and properties. *Dalton Trans.* **2013**, *42*, 16375–16386. [CrossRef] [PubMed]

22. Zhao, S.; Zheng, T.-R.; Zhang, Y.-Q.; Lv, X.-X.; Li, B.-L.; Zhang, Y. Syntheses, structures and photocatalytic properties of a series of cobalt coordination polymers based on flexible bis(triazole) and dicarboxylate ligands. *Polyhedron* **2017**, *121*, 61–69. [CrossRef]

23. Erer, H.; Yeşilel, O.Z.; Arıcı, M. A Series of zinc(II) 3D→3D interpenetrated coordination polymers based on thiophene-2,5-dicarboxylate and bis(Imidazole) derivative linkers. *Cryst. Growth Des.* **2015**, *15*, 3201–3211. [CrossRef]

24. Song, C.; Liu, Q.; Liu, W.; Cao, Z.; Ren, Y.; Zhou, Q.; Zhang, L. Two new luminescent Zn(II) coordination polymers with different interpenetrated motifs. *J. Mol. Struct.* **2015**, *1099*, 49–53. [CrossRef]

25. Sun, D.; Xu, M.-Z.; Liu, S.-S.; Yuan, S.; Lu, H.-F.; Feng, S.-Y.; Sun, D.-F. Eight Zn(II) coordination networks based on flexible 1,4-di(1H-imidazol-1-yl)butane and different dicarboxylates: Crystal structures, water clusters, and topologies. *Dalton Trans.* **2013**, *42*, 12324–12333. [CrossRef] [PubMed]

26. Xue, L.-P.; Chang, X.-H.; Ma, L.-F.; Wang, L.-Y. Four d^{10} metal coordination polymers based on bis(2-methyl imidazole) spacers: Syntheses, interpenetrating structures and photoluminescence properties. *RSC Adv.* **2014**, *4*, 60883–60890. [CrossRef]

27. Zhang, C.-Y.; Wang, M.-Y.; Li, Q.-T.; Qian, B.-H.; Yang, X.-J.; Xu, X.-Y. Hydrothermal synthesis, crystal structure, and luminescent properties of two Zinc(II) and Cadmium(II) 3D metal-organic frameworks. *Zeitschrift für Anorganische und Allgemeine Chemie* **2013**, *639*, 826–831. [CrossRef]

28. Zhang, L.; Li, X.; Zhang, Y. Two double and triple interpenetrated Cd(II) and Zn(II) coordination polymers based on mixed O- and N-donor ligands: Syntheses, crystal structures and luminescent properties. *J. Mol. Struct.* **2016**, *1103*, 56–60. [CrossRef]

29. Wang, J.; Zhu, X.; Cui, Y.-F.; Li, B.-L.; Li, H.-Y. A polythreading coordination array formed from 2D grid networks and 1D chains. *CrystEngComm* **2011**, *13*, 3342–3344. [CrossRef]

30. Liu, Y.-Y.; Li, J.; Ma, J.-F.; Ma, J.-C.; Yang, J. A series of 1D, 2D and 3D coordination polymers based on a 5-(benzonic-4-ylmethoxy)isophthalic acid: Syntheses, structures and photoluminescence. *CrystEngComm* **2012**, *14*, 169–177. [CrossRef]

31. Kan, W.-Q.; Ma, J.-F.; Liu, B.; Yang, J. A series of coordination polymers based on 5,5′-(ethane-1,2-diyl)-bis(oxy)diisophthalic acid and structurally related N-donor ligands: Syntheses, structures and properties. *CrystEngComm* **2012**, *14*, 286–299. [CrossRef]

32. Zhang, K.-L.; Hou, C.-T.; Song, J.-J.; Deng, Y.; Li, L.; Ng, S.W.; Diao, G.-W. Temperature and auxiliary ligand-controlled supramolecular assembly in a series of Zn(II)-organic frameworks: Syntheses, structures and properties. *CrystEngComm* **2012**, *14*, 590–600. [CrossRef]

33. Zhao, S.; Zhu, X.; Wang, J.; Yang, Z.; Li, B.L.; Wu, B. Two unusual 3D and 2D zinc coordination polymers containing 2D or 1D [Zn$_2$(btec)]$_n$ based on flexible bis(triazole) and rigid benzenetetracarboxylate co-ligands. *Inorg. Chem. Commun.* **2012**, *26*, 37–41. [CrossRef]

34. Zhu, X.; Yang, Y.; Jiang, N.; Li, B.; Zhou, D.; Fu, H.; Wang, N. Hydrothermal assembly of two new 3D Zinc(II) pcu nets: Coordination chemistry, crystal structures, and fluorescence properties. *Zeitschrift für Anorganische und Allgemeine Chemie* **2015**, *641*, 699–703. [CrossRef]

35. Wang, J.; Qian, X.; Cui, Y.-F.; Li, B.-L.; Li, H.-Y. Syntheses, structures, and luminescence of three 4-connected zinc coordination polymers with bis(1,2,4-triazol-1-yl)propane and benzenebiscarboxylate. *J. Coord. Chem.* **2011**, *64*, 2878–2889. [CrossRef]

36. Yan, Z.-H.; Han, L.-L.; Zhao, Y.-Q.; Li, X.-Y.; Wang, X.-P.; Wang, L.; Sun, D. Three mixed-ligand coordination networks modulated by flexible N-donor ligands: Syntheses, topological structures, and temperature-sensitive luminescence properties. *CrystEngComm* **2014**, *16*, 8747–8755. [CrossRef]

37. Tian, L.; Niu, Z.; Yang, N.; Zou, J.-Y. Crystal structures and luminescent properties of zinc(II) and cadmium(II) compounds constructed from 5-sulfoisophthalic acid and flexible bis-triazole ligands. *Inorg. Chim. Acta* **2011**, *370*, 230–235. [CrossRef]

38. Hong-Bing, Y.; Jian-Ge, W. Crystal structure of catena-[(μ₂-5-methylisophthalato)(μ₂-1,3-bis-(1,2,4-triazol-1-yl)propane)]Zinc(II)dihydrate, [Zn(C₉H₆O₄)(C₇H₁₀N₆)]·2H₂O, C₁₆H₂₀N₆O₆Zn. *Z. Krist.-New Cryst. Struct.* **2012**, *227*, 457.

39. Han, M.-L.; Chang, X.-H.; Feng, X.; Ma, L.-F.; Wang, L.-Y. Temperature and pH driven self-assembly of Zn(II) coordination polymers: Crystal structures, supramolecular isomerism, and photoluminescence. *CrystEngComm* **2014**, *16*, 1687–1695. [CrossRef]

40. Tian, L.; Yang, N.; Zhao, G. Syntheses, structures, and luminescent properties of zinc(II) complexes assembled with aromatic polycarboxylate and 1,3-bis(1,2,4-triazol-1-yl)propane. *Inorg. Chem. Commun.* **2010**, *13*, 1497–1500. [CrossRef]

41. Luo, Y.-H.; Yue, F.-X.; Yu, X.-Y.; Gu, L.-L.; Zhang, H.; Chen, X. A series of entangled ZnII/CdII coordination polymers constructed from 1,3,5-benzenetricarboxylate acid and flexible triazole ligands. *CrystEngComm* **2013**, *15*, 8116. [CrossRef]

42. Luo, Y.-H.; Tao, C.-Z.; Zhang, D.-E.; Ma, J.-J.; Liu, L.; Tong, Z.-W.; Yu, X.-Y. Three new three dimensional Zn(II)-benzenetetracarboxylate coordination polymers: Syntheses, crystal structures and luminescent properties. *Polyhedron* **2017**, *123*, 69–74. [CrossRef]

43. Semitut, E.Y.; Sukhikh, T.S.; Filatov, E.Y.; Anosova, G.A.; Ryadun, A.A.; Kovalenko, K.A.; Potapov, A.S. Synthesis, crystal structure, and luminescent properties of novel zinc metal-organic frameworks based on 1,3-Bis(1,2,4-triazol-1-yl)propane. *Cryst. Growth Des.* **2017**, *17*, 5559–5567. [CrossRef]

44. Feng, W.; Chang, R.N.; Wang, J.Y.; Yang, E.C.; Zhao, X.J. Four 1,3-bis(1,2,4-triazol-1-yl)propane-based metal complexes tuned by competitive coordination of mixed ligands: Synthesis, solid structure, and fluorescence. *J. Coord. Chem.* **2010**, *63*, 250–262. [CrossRef]

45. Zhu, X.; Liu, K.; Yang, Y.; Li, B.-L.; Zhang, Y. Syntheses and structures of three zinc coordination polymers with 1-D zigzag chain, double chain, and triple chain. *J. Coord. Chem.* **2009**, *62*, 2358–2366. [CrossRef]

46. Feng, X.; Zhou, L.L.; Shi, Z.-Q.; Shang, J.-J.; Wu, X.H.; Wang, L.-Y.; Zhou, J.G. Synthesis, crystal structure, and luminescence property of a new Zinc(II) complex with schiff-base containing triazole propane ancillary ligand. *Synth. React. Inorg. Met.-Org. Nano-Met. Chem.* **2013**, *43*, 1093–1098. [CrossRef]

47. Yin, G.; Zhang, Y.; Li, B.; Zhang, Y. Syntheses, structures and luminescent properties of a dimer and an one-dimensional chain coordination polymer with the flexible bis(triazole) and hydroxybenzoate ligands. *J. Mol. Struct.* **2007**, *837*, 263–268. [CrossRef]

48. Liu, S.-Y.; Tian, L. Poly[hemi(hexaaquazinc) [[μ₂-1,3-bis(1,2,4-triazol-1-yl)methane](μ₂-5-sulfonatobenzene-1,3-dicarboxylato)zinc] sesquihydrate]. *Acta Crystallogr. E* **2011**, *67*, m950–m951. [CrossRef] [PubMed]

49. Zhu, Y.-Y.; Zhu, M.-S.; Yin, T.-T.; Meng, Y.-S.; Wu, Z.-Q.; Zhang, Y.-Q.; Gao, S. Cobalt(II) coordination polymer exhibiting single-ion-magnet-type field-induced slow relaxation behavior. *Inorg. Chem.* **2015**, *54*, 3716–3718. [CrossRef] [PubMed]

50. McKinnon, J.J.; Jayatilaka, D.; Spackman, M.A. Towards quantitative analysis of intermolecular interactions with hirshfeld surfaces. *Chem. Commun.* **2007**, 3814–3816. [CrossRef]

51. Soliman, S.M.; El-Faham, A. Synthesis, crystal structure and hirshfeld topology analysis of polymeric Silver(I) complex with s-triazine-type ligand. *Crystals* **2017**, *7*, 160. [CrossRef]

52. *PCPDFWin, Version 1.30*, Swarthmore: Swarthmore, PA, USA, 1997.

53. *APEX2, Version 2.0, SAINT, Version 8.18c, and SADABS, Version 2.11*; Bruker Advanced X-ray Solutions; Bruker AXS Inc.: Madison, WI, USA, 2000–2012.

54. Sheldrick, G.M. Crystal structure refinement with SHELXL. *Acta Crystallogr. Sect. C* **2015**, *71*, 3–8. [CrossRef] [PubMed]

55. Van der Sluis, P.; Spek, A.L. BYPASS: An effective method for the refinement of crystal structures containing disordered solvent regions. *Acta Crystallogr. Sect. A* **1990**, *46*, 194–201. [CrossRef]

56. Turner, M.J.; McKinnon, J.J.; Wolff, S.K.; Grimwood, D.J.; Spackman, P.R.; Jayatilaka, D.; Spackman, M.A. *CrystalExplorer17*; University of Western Australia: Perth, Australia, 2017.

crystals

MDPI

Review

Perplexing Coordination Behaviour of Potentially Bridging Bipyridyl-Type Ligands in the Coordination Chemistry of Zinc and Cadmium 1,1-Dithiolate Compounds

Edward R. T. Tiekink

Research Centre for Crystalline Materials, School of Science and Technology, Sunway University,
No. 5 Jalan Universiti, Bandar Sunway 47500, Selangor Darul Ehsan, Malaysia; edwardt@sunway.edu.my;
Tel.: +60-3-7491-7181

Received: 9 December 2017; Accepted: 31 December 2017; Published: 4 January 2018

Abstract: The X-ray structural chemistry of zinc and cadmium 1,1-dithiolates (for example, xanthate, dithiophosphate and dithiocarbamate) with potentially bridging bipyridyl-type ligands (for example, 4,4'-bipyridine) is reviewed. For zinc, the xanthates and dithiophosphates uniformly form one-dimensional coordination polymers, whereas the zinc dithiocarbamates are always zero-dimensional, reflecting the exceptional chelating ability of dithiocarbamate ligands compared with xanthates and dithiophosphates. For cadmium, one-dimensional coordination polymers are usually found, reflecting the larger size of cadmium compared with zinc, but zero-dimensional aggregates are sometimes found. Steric effects associated with the 1,1-dithiolate-bound R groups are shown to influence supramolecular aggregation and, when formed, polymer topology in order to reduce steric hindrance; the nature of the bipyridyl-type ligand can also be influential. For the dithiocarbamates of both zinc and cadmium, in instances where the dithiocarbamate ligand is functionalised with hydrogen bonding potential, extended supramolecular architectures are often formed via hydrogen bonding interactions. Of particular interest is the observation that the bipyridyl-type ligands do not always bridge zinc or cadmium 1,1-dithiolates, being monodentate instead, often in the presence of hydrogen bonding. Thus, hydroxyl-O–H ⋯ N(pyridyl) hydrogen bonds are sometimes formed in preference to M←N(pyridyl) coordinate-bonds, suggesting a competition between the two modes of association.

Keywords: crystal engineering; coordination polymers; hydrogen bonding; structural chemistry; zinc; cadmium; dithiocarbamate; xanthate; dithiophosphates; unusual coordination modes

---

1. Introduction

Bioinorganic chemistry, incorporating investigations into the natural functions of metals in biology to the development of metal-based therapeutics and diagnostic agents, has been the mainstay of coordination chemistry, involving transitions metals, main group elements and lanthanides, for many decades. However, this dominance is increasingly challenged by the development of coordination polymers of various dimensions up to and including three-dimensions, that is, metal-organic framework structures. This interest is reflected in the now over 70,000 X-ray crystal structures of coordination polymers [1] included in the Cambridge Structural Database [2]. A myriad of potential applications prompt investigations into coordination polymers. While these were initially in the realm of materials science with applications relating to energy and gas storage, photo-responsive materials, catalysis, etc. [3–7], there are increasing applications of coordination polymers relevant to bioinorganic chemists. Examples include coordination polymers functioning as carriers for drug delivery, biosensors and as therapeutics themselves [8–11]. Coordination polymers are now being constructed from biologically relevant materials [12,13].

Coordination polymers are typically neutral and are constructed from a combination of charged and neutral ligands; it is noted that charged coordination polymers are also known. Increasing the functionality of the ligands enhances the probability of attaining the three-dimensional metal–organic frameworks. Charged ligands for complexation are exemplified by carboxylates derived from, for example, acetic acid, terephthalic acid (benzene-1,4-dicarboxylic acid) and trimesic acid (benzene-1,3,5-tricarboxylic acid); that is, a series with increasing bridging capacity. In the same way, potentially bridging bipyridyl-type ligands are prominent in building coordination polymers with the most prominent example being 4,4′-bipyridine. The combination of these and related building blocks have been extensively exploited by the research groups of, for example, Fujita [14,15], Yaghi [16,17] and Zaworotko [18,19].

A class of ligands closely related to carboxylate ligands is the 1,1-dithiolates, having two sulphur atoms connected to the quaternary carbon atom rather than two oxygens. Prominent examples of 1,1-dithiolate ligands include xanthates, dithiophosphates and dithiocarbamates, Figure 1. Despite the chemical similarity of the 1,1-dithiolates with carboxylate, the evaluation of the propensity of these ligands to form coordination polymers is generally lacking. This is perhaps a little surprising given that 1,1-dithiolate ligands are well known to form bridges between metal centres [20–24], especially in the binary 1,1-dithiolates of the zinc-triad elements [25]. Indeed, it was the desire to destroy this supramolecular aggregation by the addition of nitrogen-bases that was the original motivation for investigating this chemistry, as the resulting smaller aggregates were found to be more useful as synthetic precursors for chemical vapour deposition of metal sulphide nanomaterials.

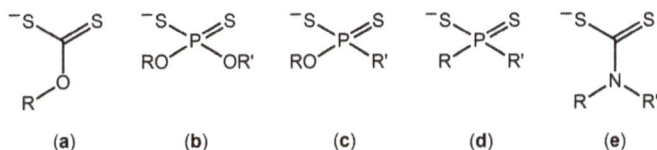

Figure 1. Chemical diagrams for (**a**) xanthate (*O*-alkyldithiocarbonate); (**b**) dithiophosphate; (**c**) dithiophosphonate; (**d**) dithiophosphinate; and (**e**) dithiocarbamate. R, R′ = alkyl, aryl.

In this bibliographic review, the structural chemistry of zinc and cadmium 1,1-dithiolates and bipyridyl-type ligands is surveyed. While one-dimensional coordination polymers are observed, there are no examples of two- let alone three-dimensional coordination polymers, at least not stabilised by bonds involving zinc or cadmium, that is, by a combination of covalent and coordinate-bonding interactions. Despite this, an interesting structural chemistry is frequently revealed. For example, often in the presence of hydrogen bonding functionality in the 1,1-dithiolate ligands, putative coordinate-bonding between zinc or cadmium and the bipyridyl-type ligand is suppressed to allow for the formation of competing hydrogen bonding interactions.

2. Methodology and Organisation

The structures included in this survey were extracted in the form of crystallographic information files (CIFs) from the Cambridge Structural Database (CSD, 2017 release with updates) [2]. Data were analysed using PLATON [26] and diagrams drawn with the graphics program DIAMOND [27]. The review is divided into a discussion of zinc structures followed by those of cadmium. Within in each category, xanthate species are discussed before dithiophosphates (and lesser known derivatives) and then dithiocarbamates. A diverse variety of bipyridyl-type ligands are represented among the structures and their chemical diagrams with abbreviations are shown in Figure 2. As a general principal, smaller bipyridyl-type ligands are discussed before larger/more complicated analogues and for each type of bipyridyl-type ligand, the molecules/aggregates are described in terms of the bulk of the R/R′ substituents of the 1,1-dithiolate ligand, with smaller groups discussed first. The discussion is focussed upon a general description of the modes of coordination of the 1,1-dithiolate and bipyridyl-type ligands and describing the resultant aggregation patterns. Details of geometric parameters are available in the original publication and are not generally considered. Similarly, generally descriptions of molecular packing are avoided unless relevant to the putative competition between hydrogen bonding and M←N(pyridyl) coordinate-bonds. In all, there are 57 zinc-containing structures and 31 cadmium analogues.

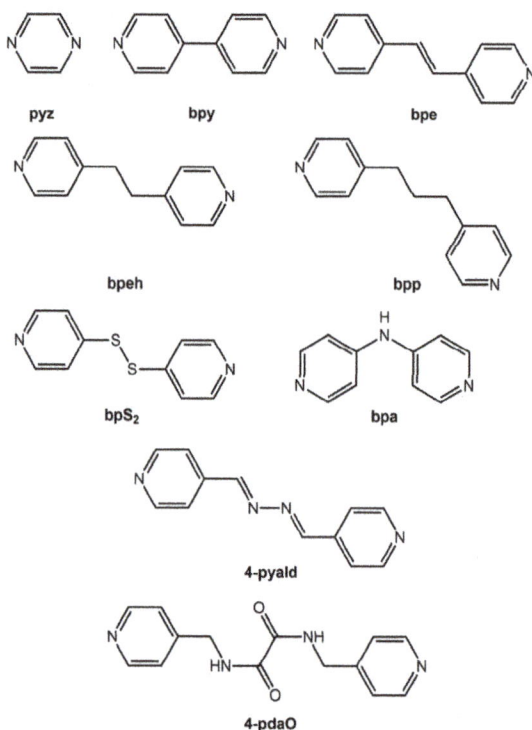

Figure 2. Chemical diagrams for bipyridyl-type ligands discussed in this review and their abbreviations: (**pyz**) pyrazine, (**bpy**) 4,4′-bipyridine, (**bpe**) *trans*-1,2-bis(4-pyridyl)ethylene, (**bpeh**) *trans*-1,2-bis(4-pyridyl)ethane, (**bpp**) *trans*-1,2-bis(4-pyridyl)propane, (**bpS₂**) bis(4-pyridyl)disulphide, (**bpa**) bis(4-pyridyl)amine, (**4-pyald**) 1,2-bis(4-pyridylmethylene)hydrazine and (**4-pdaO**) N,N′-bis (pyridin-4-ylmethyl)ethanediamide. Abbreviations for isomers and derivatives of the above: **2-bpe**, 2-pyridyl isomer of bpe; **3-pyald** and **2-pyald**, 3- and 2-pyridyl isomers of 4-pyald; **3-pdaO**, 3-pyridyl isomer of 4-pdaO; **3-pdaOt**, hydroxyl-imine tautomer of 3-pdaO; **3-pdaS**, thio analogue of 3-pdaO.

3. Discussion

3.1. Zinc Xanthate Structures

There are six structures in this category, **1–6** [28–31], and general features of these are summarised in Table 1 along with the other zinc structures, **1–57** [28–66], included in this survey. The first structure to be described contains the smallest bipyridyl-type ligand, namely pyrazine (pyr), and is a one-dimensional coordination polymer formulated as [Zn(S₂COEt)₂(pyr)]ₙ (**1**) [28]. In **1**, Figure 1a, the zinc atom lies on a special position 2/m and the pyr ligand lies about a two-fold axis with the nitrogen atoms lying on the axis. The xanthate ligands are S,S-chelating and the resulting *trans*-N$_2$S$_4$ donor set defines a distorted octahedral geometry. The topology of the chain is strictly linear owing to the aforementioned symmetry. It is apposite here to comment on the structure of the parent molecule found in **1**; that is, [Zn(S$_2$COEt)$_2$]$_n$ [67,68]. Here, each xanthate ligand is bidentate bridging and the resulting two-dimensional structure is conveniently described as being constructed by an infinite pattern of edge-shared squares, with each square comprising zinc atoms and with pairs of adjacent zinc atoms being bridged by a xanthate ligand resulting in ZnS$_4$ tetrahedra. With the addition of base, pillaring of the layers into a three-dimensional architecture might be envisaged, with suitable adjustments in the coordination geometries about the zinc atoms. Instead, the original two-dimensional architecture of [Zn(S$_2$COEt)$_2$]$_n$ is disrupted in **1** to form a one-dimensional aggregation pattern. This principle of perturbing the supramolecular architecture observed in the parent 1,1-dithiolate structure is generally applicable to the structures described herein.

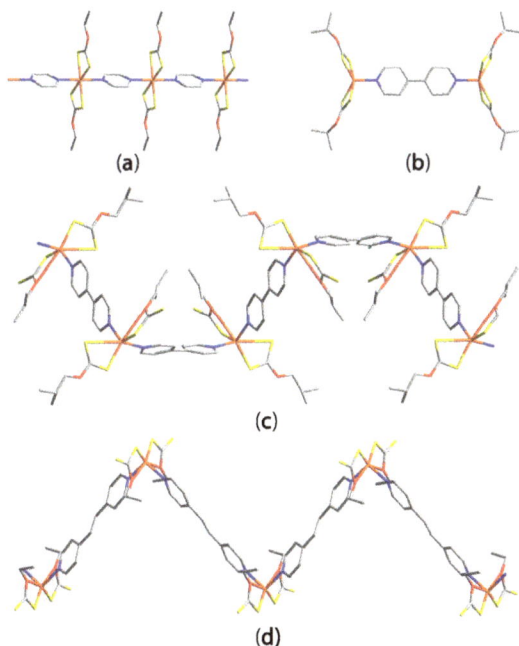

Figure 3. Aggregation in the structures of (**a**) [Zn(S$_2$COEt)$_2$(pyr)]$_n$ (**1**); (**b**) [Zn(S$_2$COiPr)$_2$]$_2$(bpy) (**2**); (**c**) [Zn(S$_2$COiBu)$_2$(bpy)]$_n$ (**3**); and (**d**) [Zn(S$_2$COEt)$_2$(bpe)]$_n$ (**4**). Colour code: zinc or cadmium, orange; sulphur, yellow; oxygen, red, nitrogen, blue; carbon, grey. Non-acidic hydrogen atoms and non-participating species, typically solvent molecules, have been omitted in all diagrams.

Table 1. Summary of the general features of zinc structures **1–57** [28–66].

Compound	R/R′	N-Ligand	Donor Set	Motif	Ref.
1	Et	pyr	N$_2$S$_4$	linear chain	[28]
2[1]	iPr	bpy	NS$_4$	dimer	[29]
3	iBu	bpy	N$_2$OS$_3$	helical chain	[30]
4	Et	bpe	N$_2$O$_2$S$_2$	zig-zag chain	[31]
5	nBu	bpe	N$_2$O$_2$S$_2$	zig-zag chain	[31]
6[2]	Cy	bpe	NS$_4$	dimer	[31]
7	iPr	pyr	NS$_3$+S	dimer	[32]

Table 1. *Cont.*

Compound	R/R′	N-Ligand	Donor Set	Motif	Ref.
8	Et	bpy	N_2S_2 N_2OS_2	zig-zag chain	[33]
9	iPr	bpy	N_2S_2	zig-zag chain	[34]
10	Cy	bpy	NS_4	dimer	[35]
11	iPr	bpe	N_2S_2	zig-zag chain	[36]
12	iBu	bpe	N_2S_4	linear chain	[37]
13	Cy	bpe	N_2S_4	linear chain	[35]
14 [3]	iPr	bpeh	NS_4	dimer	[38]
15	Et	bpeh	N_2S_2	zig-zag chain	[35]
16 [4]	iPr	bpeh	N_2S_2	zig-zag chain	[35]
17	iBu	bpeh	N_2S_4	linear chain	[39]
18	Cy	bpeh	N_2S_2	zig-zag chain	[35]
19	iPr	bpp	N_2S_2	helical chain	[32]
20	iPr	bpS_2	N_2OS_2	zig-zag chain	[32]
21	iPr	bpa	N_2S_2	zig-zag chain	[32]
22	Cy	4-pyald	N_2S_4	zig-zag chain	[40]
23	iPr	2-bpe	NS_4	dimer	[32]
24	Cy	2-bpe	NS_3	monomer	[32]
25	iPr	3-pyald	N_2S_4	linear chain	[41]
26	Me, 4-MePh	bpeh	N_2S_2	zig-zag chain	[42]
27	iBu	bpy	N_2S_4	linear	[43]
28	nPr	bpy	NS_4	monomer	[44]
29	Me	bpy	NS_4	dimer	[45]
30	Et	bpy	NS_4	dimer	[49,50]
31	nPr	bpy	NS_4	dimer	[44]
32 [5]	iPr	bpy	NS_4	dimer	[48]
33	iBu	bpy	NS_4	dimer	[49]
34	CH_2Ph	bpy	NS_4	dimer	[50]
35 [6]	CH_2CH_2	bpy	NS_4	dimer	[51]
36 [7]	CH_2CH_2	bpy	NS_4	dimer	[52]
37	Me	bpe	NS_4	dimer	[53]
38 [8]	Et	bpe	NS_4	dimer	[54]
39	iPr	bpe	NS_4	dimer	[55]
40 [9]	Et	bpe	NS_4	dimer	[56]
41	nPr	bpeh	NS_4	dimer	[57]
42	Et	4-pdaO	NS_4	dimer	[58]
43 [10]	Me	3-pdaO	NS_4	dimer	[59]
44	nPr	3-pdaO	NS_4	dimer	[59]
45	CH_2CH_2OH	pyr	NS_4	dimer	[60]
46 [11]	CH_2CH_2OH	pyr	NS_4	dimer	[60]
47	Me, CH_2CH_2OH	bpy	NS_4	dimer	[61]
48	Et, CH_2CH_2OH,	bpy	NS_4	dimer	[61]
49 [12]	Et, CH_2CH_2OH	bpy	NS_4	dimer	[61]
50	CH_2CH_2OH	bpy	NS_4	dimer	[61]
51 [13]	iPr, CH_2CH_2OH	bpy	NS_4	monomer	[62]
52	Me, CH_2CH_2OH	4-pyald	NS_4	monomer	[63]
53	Me, CH_2CH_2OH	3-pdaO	NS_4	dimer	[64]
54	CH_2CH_2OH	3-pdaOt	NS_4	dimer	[64]
55 [10]	Me, CH_2CH_2OH	3-pdaS	NS_4	dimer	[65]
56 [14]	Me, CH_2CH_2OH	3-pdaS	NS_4	dimer	[65]
57 [15]	Me, CH_2CH_2OH	triazine	N_3S_2	monomer	[66]

[1] Dichloromethane mono-solvate; [2] Chloroform di-solvate; [3] Acetonitrile di-solvate; [4] Chloroform hemi-solvate; [5] Toluene di-solvate; [6] NR_2 = pyrrolidine; [7] NR_2 = piperidine; [8] Chloroform mono-solvate; [9] Co-crystallised with one uncoordinated bpe molecule; [10] Dimethylformamide di-solvate; [11] Dioxane di-solvate; [12] Methanol di-solvate; [13] Co-crystallised with half an uncoordinated bpy molecule; [14] Co-crystallised with two S_8 molecules molecule; [15] Dioxane sesqui-solvate.

The next structures to be described feature the 4,4'-bipyridine (bpy) ligand which is well represented in the present survey. The structure of [Zn(S$_2$COiPr)$_2$]$_2$(bpy) (**2**) [29] is shown in Figure 1b and is binuclear; the molecule has two-fold symmetry with the zinc atoms lying on the axis. The xanthate ligands are S,S-chelating and the five-coordinate geometries are completed by a pyridyl-N atom. The NS$_4$ donor set is heavily distorted with the value of τ computing to 0.63, which compares to 1.0 for an ideal trigonal-bipyramid and 0.0 for an ideal square-pyramid [69]. The second zinc xanthate structure with bpy is a coordination polymer as the ratio of components is 1:1 rather than 2:1 as in **2**. In [Zn(S$_2$COiBu)$_2$(bpy)]$_n$ (**3**) [30], one xanthate ligand is S,S-chelating but with disparate Zn–S bond lengths while the other is S,O-chelating with the Zn–O separation being long at 3.11 Å. While comparatively rare, such S,O-chelating modes are known in the structural chemistry of metal xanthates [20]. In **3**, the resulting N$_2$OS$_3$ coordination geometry is based on a distorted octahedron. The topology of the chain is helical, Figure 1c. A difference in the conformation of the bpy ligand in the structures of **2** and **3** is noted, namely the dihedral angle between the pyridyl rings in **2** is 13.9° compared with 29.0° in **3**, indicating conformational flexibility. The three remaining structures in this section contain the *trans*-1,2-bis(4-pyridyl)ethylene (bpe) ligand in a bridging mode [31]. In [Zn(S$_2$COR)$_2$(bpe)]$_n$, for R = Et (**4**) and nBu (**5**), both xanthate ligands are S,O-chelating resulting in N$_2$O$_2$S$_2$ donor sets which are best described as being skew-trapezoidal bipyramidal in which the pyridyl-N atoms lie over the relatively long Zn–O bonds. The coordination polymer in each of **4** and **5** has a zig-zag topology as illustrated for **4** in Figure 1d. When the steric bulk of the R group in the xanthate ligand is increased, a new aggregation pattern in observed. In [Zn(S$_2$COCy)$_2$]$_2$(bpe) (**6**) [31], isolated from analogous 1:1 solutions that yielded **4** and **5**, the xanthate ligands revert to S,S-coordination modes and a binuclear aggregate is formed, similar to that shown in Figure 1b. Such steric effects in the coordination chemistry of metal 1,1-thiolates is well documented [25,68,70]. With τ = 0.51, the NS$_4$ coordination geometry in **6** is almost exactly intermediate between the two ideal five-coordinate geometries. The influence of aggregation patterns in **4**–**6**—that is, polymeric versus dimeric—was shown to have a distinct influence upon the solid-state luminescence responses [31].

Although only including six structures, the foregoing offers a microcosm of the variability of structures covered in this review. Thus, different coordination modes of the xanthate ligand are evident leading to distinct coordination geometries, different aggregation patterns, sometimes related to the steric bulk of the substituents, and, when formed variable topologies of coordination polymers and conformational variations in the bridging ligands.

3.2. Zinc Dithiophosphate and Related Structures

The first dithiophosphate structure to be described in this section is a centrosymmetric binuclear species similar to that observed for **2**, Figure 1b. In {Zn[S$_2$P(OiPr)$_2$]$_2$}$_2$(pyr) (**7**) [32], one dithiophosphate ligand is chelating while the other, with Zn–S bond lengths differing by nearly 1 Å, is coordinating in an asymmetric mode or taken to an extreme, is monodentate. The NS$_3$ donor set is distorted tetrahedral with the widest angle at the zinc atom of approximately 132° correlating with the close approach of the non-coordinating sulphur atom. Crystals of **7** were isolated from solutions containing a 1:1 ratio of reagents but only the 2:1 species was isolated [32]. As evident from Figure 4a, the environment about the binuclear is congested and steric hindrance is likely the cause for the lack of polymer formation.

Compounds **8**–**10** involve bridging bpy ligands. In {Zn[S$_2$P(OEt)$_2$]$_2$(bpy)}$_n$ (**8**) [33], two independent formula units comprise the asymmetric unit. One of the independent zinc atoms is coordinated within a distorted tetrahedral N$_2$S$_2$ donor set; both dithiophosphate ligands coordinate in the monodentate mode. By contrast, the second independent zinc atom is five coordinate as one of the dithiophosphate ligands is S,O-coordinating. The NO$_2$S$_2$ donor set is intermediate between ideal trigonal-bipyramidal and square-pyramidal geometries (τ = 0.53). There are three independent bpy molecules in the structure with two located about a centre of inversion. For these, the twist between rings is 0°. For the bpy molecule bridging the independent zinc atoms, the twist is 29°. The resultant coordination polymer has a zig-zag topology as shown in Figure 4b. To a first approximation, a similar coordination polymer is found in the crystal of {Zn[S$_2$P(OiPr)$_2$]$_2$(bpy)}$_n$ (**9**) [34]. The asymmetric unit of **9** comprises four independent repeat units but, each zinc atom exists within a distorted tetrahedral N$_2$S$_2$ donor set. The dihedral angles for the bpy ligands range from 26 to 45°. The third structure has bulky cyclohexyl substituents and therefore, owing to steric hindrance, only binuclear {Zn[S$_2$P(OCy$_2$)$_2$]$_2$}$_2$(bpy) (**10**) [35] could be isolated. Here, the dithiophosphate ligands coordinate in a S,S-mode leading to a NS$_4$ donor set which tends towards a square-pyramidal geometry with the pyridyl-N in the apical position (τ = 0.44). The steric crowding in the centrosymmetric molecule is apparent from Figure 4c. There are also three structures with bridging bpe ligands. The structure of {Zn[S$_2$P(OiPr)$_2$]$_2$(bpe)}$_n$ (**11**) [36] with a N$_2$S$_2$ donor sets resembles the zig-zag polymeric structure of **10**. However, for {Zn[S$_2$P(OR)$_2$]$_2$(bpe)}$_n$, for R = iBu (**12**) [37] and R = Cy (**13**) [35], distorted octahedral geometries based on *trans*-N$_2$S$_4$ donor sets are observed leading to effectively linear coordination polymers. Particularly noteworthy is the polymeric structure of **13**, the first thus far in this survey featuring bulky cyclohexyl groups. As seen from Figure 4d, the ethene link between the 4-pyridyl rings introduces more space between adjacent zinc centres which readily accommodates the large cyclohexyl rings and thereby, enables the formation of a polymeric chain.

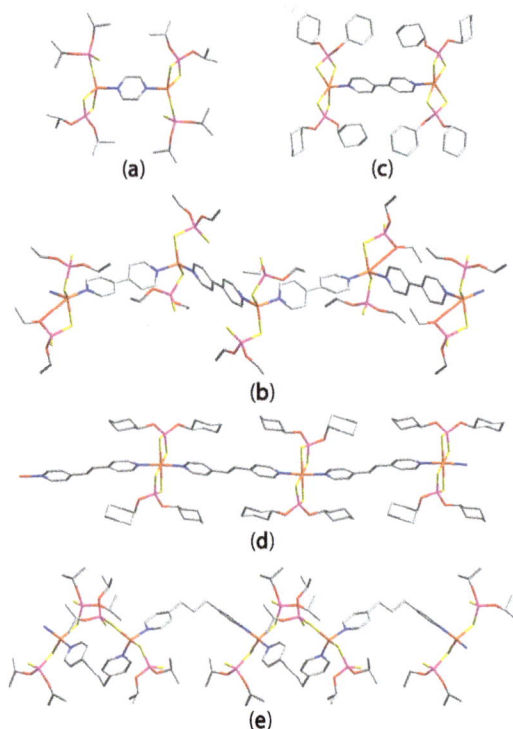

Figure 4. Aggregation in the structures of (**a**) {Zn[S$_2$P(OiPr)$_2$]$_2$}$_2$(pyr) (**7**); (**b**) {Zn[S$_2$P(OEt)$_2$]$_2$(bpy)}$_n$ (**8**); (**c**) {Zn[S$_2$P(OCy$_2$)$_2$]$_2$}$_2$(bpy) (**10**); (**d**) {Zn[S$_2$P(OCy)$_2$]$_2$(bpe)}$_n$ (**13**); and (**e**) [Zn(S$_2$COiPr)$_2$(bpp)]$_n$ (**19**). Additional colour code: phosphorus, pink.

The next five structures feature the saturated version of the bpe ligand, that is, *trans*-1,2-bis (4-pyridyl)ethane (bpeh), and it is particularly notable that both 2:1 and 1:1 structures were characterised during systematic studies of these molecules. Thus, {Zn[S$_2$P(OiPr$_2$)$_2$]$_2$}$_2$(bpeh) (**14**) [38] is centrosymmetric and resembles the aggregate shown in Figure 4c. A difference arises in that the NS$_4$ coordination geometry tends towards trigonal-bipyramidal (τ = 0.58). In the same way, the structures

of {Zn[S$_2$P(OR)$_2$]$_2$(bpeh)}$_n$, for R = Et (**15**) [35] and R = iPr (**16**) [35], the 1:1 analogue of **14**, resemble
the zig-zag polymer chains illustrated in Figure 4b. Continuing this theme, {Zn[S$_2$P(OiBu)$_2$]$_2$(bpeh)}$_n$
(**17**) [39] mimics the linear chain and *trans*-N$_2$S$_4$ coordination geometry shown for **12** in Figure 4d.
A difference between the bpe- and bpeh-containing structures occurs in the case when R = Cy. Thus,
in {Zn[S$_2$P(OCy)$_2$]$_2$(bpeh)}$_n$ (**18**) [35], the topology of the chain is zig-zag in contrast to the linear chain
found for **13**. This suggests a subtle interplay between the adoption of linear versus zig-zag chains.
It is noted the Zn $\cdots$ Zn separation in the bpe-containing structure, **13**, is 13.8 Å compared with 13.2 Å
in **18**. Thus, a more compact arrangement is found for the zig-zag chain. This is achievable in the case
when the bridging ligand is bpeh, as the kink in the molecule alleviates steric hindrance.

There is a single example of a structure with a bridging *trans*-1,2-bis(4-pyridyl)propane (bpp)
ligand, namely {Zn[S$_2$P(OiPr)$_2$]$_2$(bpp)}$_n$ (**19**) [32]. Each of the two independent zinc atoms in the
structure have distorted tetrahedral coordination geometries and, owing to the curved nature of the
n-propyl bridge between the 4-pyridyl residues, the resulting coordination polymer has a helical
topology as shown in Figure 4e. The following structures to be summarised contain less frequently
studied bipyridyl-type molecules.

The ligand, bis(4-pyridyl)disulphide (bpS$_2$), is bridging in {Zn[S$_2$P(OiPr)$_2$]$_2$(bpS$_2$)}$_n$ (**20**) [32].
In **20**, one dithiophosphate ligand is S-monodentate and the other is S,O-chelating but with
the 2n–O separation about 0.9 Å longer than the Zn–S bond length. The N$_2$OS$_2$ donor set defines
a coordination geometry tending towards square-pyramidal with the value of τ being 0.38. The resultant
coordination polymer has a zig-zag topology (glide symmetry) and is illustrated in Figure 5a.
A zig-zag coordination polymer is also found in the crystal of {Zn[S$_2$P(OiPr)$_2$]$_2$(bpa)}$_n$ (**21**) [32],
where bpa is bis(4-pyridyl)amine; the zinc atom exists in a distorted tetrahedral N$_2$S$_2$ donor set. The
notable characteristic of the bpa ligand is the presence of hydrogen bonding functionality. Indeed,
amine-N–H $\cdots$ O(alkoxy) hydrogen bonds are formed which serve to link chains into supramolecular
layers as shown in Figure 5b.

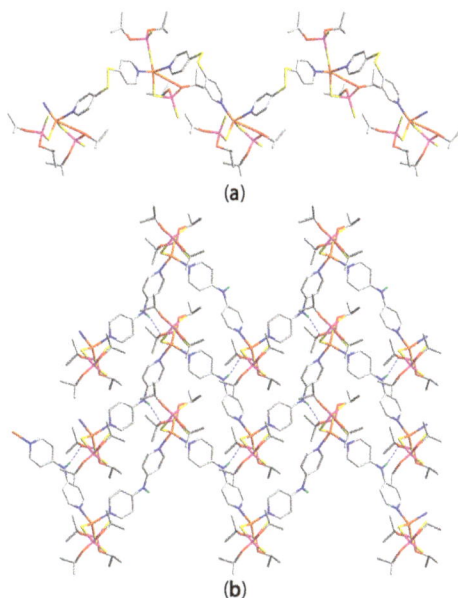

(a)

(b)

Figure 5. Aggregation in the structures of (**a**) {Zn[S$_2$P(OiPr)$_2$]$_2$(bpS$_2$)}$_n$ (**20**); and (**b**) {Zn[S$_2$P(OiPr)$_2$]$_2$(bpa)}$_n$
(**21**). Additional colour code: hydrogen, green. The amine-N–H $\cdots$ O(alkoxy) hydrogen bonds in (**b**)
are shown as blue dashed lines.

The 1,2-bis(4-pyridylmethylene)hydrazine (4-pyald; 4-pyridylaldazine) ligand is the first example of a bipyridyl-type ligand in this survey with heteroatoms in the bridge linking the 4-pyridyl residues. With four atoms in the link—that is, an all-*trans* CN_2C sequence, separating the 4-pyridyl residues—there is sufficient space to accommodate the bulky cyclohexyl groups in {Zn[S_2P(OCy)$_2$]$_2$(4-pyald)}$_n$ (**22**) [40], and a zig-zag coordination polymer ensues, Figure 6a. The curious feature of this structure relates to the relative disposition of the sulphur atoms. In the related preceding structures, *trans*-N_2S_4 donor sets were noted but, here the donor set is *cis*-N_2S_4.

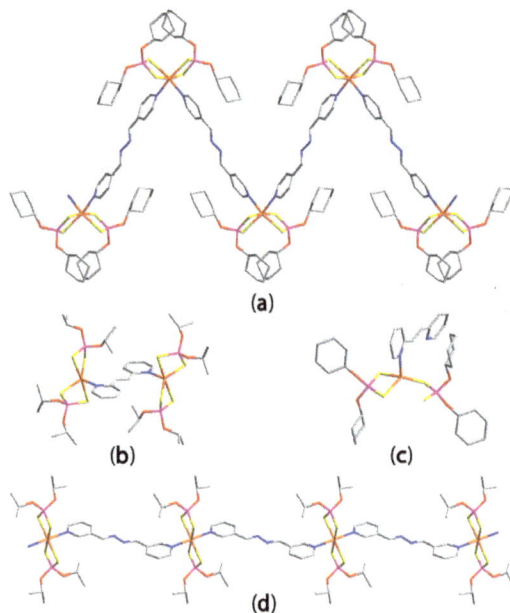

Figure 6. Aggregation in the structures of (a) {Zn[S_2P(OCy)$_2$]$_2$(4-pyald)}$_n$ (**22**); (b) {Zn[S_2P(OiPr)$_2$]$_2$(2-bpe)}$_2$ (**23**); (c) Zn[S_2P(OCy)$_2$]$_2$(2-bpe) (**24**); and (d) {Zn[S_2P(OiPr)$_2$]$_2$(3-pyald)}$_n$ (**25**).

Thus far, all bipyridyl-type ligands described have had the 4-pyridyl isomer; in the three remaining dithiophosphate structures, isomeric 2-pyridyl and 3-pyridyl species connect zinc atoms. Unlike the polymeric bpe analogue, **11**, the structure of Zn[S_2P(OiPr)$_2$]$_2$(2-bpe)}$_2$ (**23**) [32], where 2-bpe is the 2-pyridyl isomer of bpe, that is *trans*-1,2-bis(4-pyridyl)ethylene, is dimeric, no doubt owing to steric congestion, Figure 6b. In the centrosymmetric molecule, the NS_4 donor set approximates a trigonal-bipyramidal geometry ($\tau = 0.80$) with the less tightly bound sulphur atoms of the asymmetrically coordinating dithiophosphate ligands occupying axial positions (S–Zn–S = 177°). Increasing the steric bulk of the alkoxyl substituents to cyclohexyl means that not even dimerisation can occur. As shown in Figure 6c, the 2-bpe ligand adopts a very rare monodentate coordination mode in the monomeric structure of Zn[S_2P(OCy)$_2$]$_2$(2-bpe) (**24**) [32]. The 1,2-bis(3-pyridylmethylene)hydrazine (3-pyald) ligand is the 3-pyridyl isomer of 4-pyald seen in the structure of **22**. In {Zn[S_2P(OiPr)$_2$]$_2$(3-pyald)}$_n$ (**25**) [41], bridging is observed and a linear coordination polymer is formed, Figure 6d. The distorted *trans*-N_2S_4 coordination geometry is distorted octahedral.

There are single examples each of dithiophosphate analogues where the alkoxy groups are systematically substituted by alkyl groups, that is, giving dithiophosphonate and dithiophosphinate, Figure 1. In {Zn[S_2P(4-MePh)OMe]$_2$(bpeh)}$_n$ (**26**) [42], a tetrahedral N_2S_2 donor set and zig-zag coordination polymer is seen as in **15**, **16** and **18**. Finally, in {Zn[S_2P(iBu)$_2$]$_2$(bpy)}$_n$ (**27**) [43], an octahedral N_2S_4 donor set and linear coordination polymer is observed. This is in contrast to the zig-zag polymers observed in the bpy analogues **8** and **9** but, in accord with the linear polymers seen in the bpe (**12**) and bpeh (**17**) analogues perhaps suggesting a propensity for this type of polymer topology for isobutyl substituents.

3.3. Zinc Dithiocarbamate Structures

There are 30 dithiocarbamate structures, that is, **28–57** [44–66], to be described in this section, Table 1. While the majority of the structures conform to already-observed motifs, there are a number of unexpected modes of coordination for the bipyridyl-type ligands, often in the presence of dithiocarbamate ligands functionalised with groups capable of forming hydrogen bonding interactions. Hence, this section is divided into discussions of structures not capable of forming hydrogen bonds and those that are.

3.3.1. Zinc Dithiocarbamate Structures Not Capable of Forming Hydrogen Bonds

The most unexpected structure to be described in this section is monomeric $Zn(S_2CNEt_2)_2(bpy)$ (**28**) [44] as, rather than bridging, the bpy molecule coordinates in the monodentate mode: see Figure 7a. With a value of $\tau = 0.51$, the NS_4 donor set is nearly perfectly intermediate between square-pyramidal and trigonal-pyramidal. The remaining structures involving bpy, **29–36** [45–52] have this ligand in the expected bridging mode: see Table 1. The prototype structure of the majority of these $[Zn(S_2CNR_2)_2]_2(bpy)$ structures is the R = Me derivative, **29** [45], shown in Figure 7b. The molecule in **29** is located about a centre of inversion indicating the dihedral angle between the pyridyl rings of bpy is 0°. Identical symmetry is found in the R = nPr (**31**) [44], iPr (**32**) [48], iBu (**33**) [49] and CH_2Ph (**34**) [50] structures. There are three variations. In $\{Zn[S_2CN(CH_2)_4]_2\}_2(bpy)$ (**35**) [51], the molecule is situated about a site of symmetry 222. In $[Zn(S_2CNEt_2)_2]_2(bpy)$ (**30**) [46,47], there are two independent molecules in the asymmetric unit: one where the ZnNNZn vector is coincident with the two-fold axis, and the other where the two-fold axis is perpendicular to the ZnNNZn vector. In a sense, the structure of $\{Zn[S_2CN(CH_2)_5]_2\}_2(bpy)$ (**36**) [52] represents a synthesis of the above in that there are two independent molecules, one situated about a centre of inversion and the other molecule having 2-fold symmetry coincident with the ZnNNZn axis. In the non-centrosymmetric molecules, the range of dihedral angles between the pyridyl rings of bpy is 25 to 39° for the molecule with two-fold axes coincident with the ZnNNZn vector in **30** and the molecule in 222-symmetric **35**, respectively. In each of **29–36**, the donor set about the zinc atom is defined by NS_4 atoms. The coordination geometries range from almost square-pyramidal for the R = nPr (**31**) derivative with $\tau = 0.11$ to approaching trigonal-bipyramidal for the R = CH_2Ph (**34**) derivative with $\tau = 0.70$.

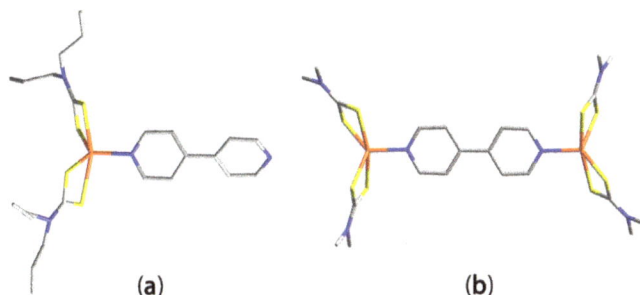

(a)　　　　　　　　　　**(b)**

Figure 7. Aggregation in the structures of (a) $Zn(S_2CNEt_2)_2(bpy)$ (**28**) and (b) $[Zn(S_2CNMe_2)_2]_2(bpy)$ (**29**).

The four zinc dithiocarbamate species with bpe, that is, **37–40** [53–56], Table 1, also adopt the same dimeric motif as illustrated in Figure 7a; the bpe is planar or close to planar in each case. The NS_4 donor sets tend to be closer to square-pyramidal in three examples, that is τ ranges from 0.13 for the R = Et **(40)** to 0.39 for R = Me **(37)** derivatives, while NS₄ is closer to trigonal-bipyramidal for the R = Et **(38)** species, with τ = 0.56. While no molecular symmetry is event for **37–39**, the binuclear molecule in **40** is disposed about a centre of inversion. The structure of **40** is of particular interest as it was isolated as a 1:1 lattice adduct with a non-coordinating molecule of bpe during unsuccessful attempts to encourage polymer formation by having an excess of bpe in solution during crystallisation [56]. The final structure in this section also follows the motif shown in Figure 7b, that is, binuclear {Zn[S₂CN(nPr)₂]₂}₂(bpeh) **(41)** [57]. The molecule has no symmetry and the NS₄ donor set approaches trigonal-bipyramidal with τ values of 0.64 and 0.72 for the independent zinc atoms.

3.3.2. Zinc Dithiocarbamate Structures Capable of Forming Hydrogen Bonds

Many of the structures described in this section conform to the motifs established earlier in this review. However, with hydrogen bonding functionality, supramolecular aggregation between the aggregates usually occurs leading to supramolecular architectures of higher dimensionality.

The first three structures to be discussed in this category have dialkyldithiocarbamate ligands coupled with bipyridyl-type ligands that have hydrogen bonding potential, for example *N*,*N'*-bis(pyridin-4-ylmethyl)ethanediamide (4-pdaO). In [Zn(S₂CNEt₂)₂]₂(4-pdaO) **(42)** [58], the binuclear molecule is disposed about a centre of inversion with the crank-shaft shape of the 4-pdaO bridge leading to the distinctive kink in the molecule, Figure 8a. The NS₄ donor set approximates a square-pyramid as indicated by the value of τ = 0.20. The next two molecules contain the 3-pyridyl isomer of 4-pdaO, *N*,*N'*-bis(pyridin-3-ylmethyl)ethanediamide (3-pdaO) and two distinctive conformations are observed. In [Zn(S₂CNMe₂)₂]₂(3-pdaO) **(43)** [59], isolated as a di-dimethylformamide (DMF) solvate, a similar centrosymmetric conformation for the 3-pdaO molecule is seen as for **42**. Similarly, the value of τ = 18 indicates the NS₄ coordination geometry is based on a square-pyramid. The co-crystallised DMF molecules are associated with the binuclear molecule via amide-N–H ⋯ O(DMF) hydrogen bonds, Figure 8b, so no further supramolecular association is possible in the crystals, at least via hydrogen bonding. A major conformational change is evident for the 3-pdaO molecule in {Zn[S₂CN(nPr₂)]₂}₂(3-pdaO) **(44)** [59]. As seen from Figure 8c, the bridging 3-pdaO molecule has a U-shape; the molecule has 2-fold symmetry. In contrast to **42**, the NS₄ coordination geometry in **43** is based on a trigonal-pyramid with τ = 0.76. While such remarkable differences in conformation may seem incongruent, computational chemistry performed on 3-pdaO show that the energy difference between the extended form, where the 3-pyridyl rings are anti-periplanar, and the curved form, where the rings are syn-periplanar, is less than 1 kcal/mol [71]. This observation highlights the conformational flexibility of this class of molecule and their related thioamide derivatives [72]. A consequence of the U-shape of the molecule in **43** is that rows of alternating oppositely orientated molecules are linked by amide-N–H ⋯ O(amide) hydrogen bonds leading to a supramolecular tape, as in Figure 8d.

The structure of {Zn[S₂CN(CH₂CH₂OH)₂]₂}₂(pyr) has been determined solvent-free **(45)**, as well as its di-dioxane solvate **(46)** [60]. The molecular structures resemble the binuclear aggregate shown in Figure 7b. In **45**, the molecule is disposed about a centre of inversion and the same is true for each of the two independent molecules in **46**. The values of τ value from 0.19 **(45)** to 0.26 and 0.40 **(46)** indicating a tendency towards square-pyramidal. In each of **45** and **46** there is extensive hydroxyl-O–H ⋯ O(hydroxyl) hydrogen bonding, involved all hydroxyl groups as acceptors and donors, leading to a three-dimensional architecture, Figure 9. As discussed in the original publication, there are no solvent accessible voids in **45** and the key crystal indicators, that is density (1.709 compared with 1.570 g/cm³) and packing efficiency (74.9 compared with 71.6%), suggest that **46** represents a less efficient packing arrangement, owing to the presence of dioxane molecules. Consistent with this is the lack of strong intermolecular interactions between the host lattice and the dioxane molecules resident in the channels. The rationale for the formation of **46** rests with the observation that owing to the high boiling point of dioxane, the dioxane molecules were trapped in the crystal under the conditions of specific crystallisation experiment [60].

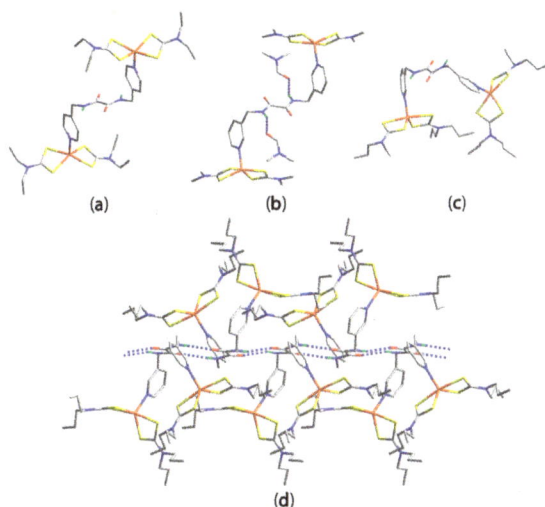

Figure 8. Aggregation in the structures of (**a**) [Zn(S$_2$CNEt)$_2$]$_2$(4-pdaO) (**42**); (**b**) [Zn(S$_2$CNMe)$_2$]$_2$ (3-pdaO).2-di-dimethylformamide (DMF) (**43**); (**c**) {Zn[S$_2$CN(nPr$_2$)]$_2$}$_2$(3-pdaO) (**44**); and (**d**) supramolecular tape mediated by amide-N–H⋯O(amide) hydrogen bonds (blue dashed lines) in **44**.

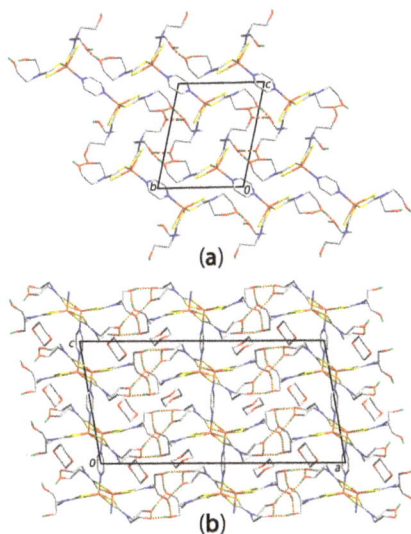

Figure 9. Three-dimensional supramolecular aggregation in the structures of (**a**) {Zn[S$_2$CN (CH$_2$CH$_2$OH)$_2$]$_2$}$_2$(pyr) (**45**); and (**b**) {Zn[S$_2$CN(CH$_2$CH$_2$OH)$_2$]$_2$}$_2$(pyr).2dioxane (**46**). The hydroxyl-O–H⋯O(hydroxyl) hydrogen bonds are shown as orange dashed lines.

The next four structures, having bridging bpy ligands, resemble the molecule shown in Figure 7b. As established above, there are no apparent patterns in the symmetry of the molecules nor in the adoption of coordination geometry. In {Zn[S$_2$CN(Me)CH$_2$CH$_2$OH]$_2$}$_2$(bpy) (**47**) [61], the molecule is disposed about a centre of inversion and the zinc atom has an intermediate coordination geometry between the two extremes ($\tau = 0.54$). By contrast, the ethyl analogue, {Zn[S$_2$CN(Et)CH$_2$CH$_2$OH]$_2$}$_2$(bpy) (**48**) [61], is non-symmetric with coordination geometries close to square-pyramidal ($\tau = 0.21$) and intermediate between this and trigonal-bipyramidal ($\tau = 0.51$) for the second zinc atom. However, the di-methanol solvate of **48**, Zn[S$_2$CN(Et)CH$_2$CH$_2$OH]$_2$}$_2$(bpy).2CH$_3$OH (**49**) [61], is centrosymmetric with a coordination geometry approaching trigonal-bipyramidal ($\tau = 0.62$). In the fourth structure of this series, with a symmetrically substituted dithiocarbamate ligand, {Zn[S$_2$CN(CH$_2$CH$_2$OH)$_2$]$_2$}$_2$(bpy) (**50**) [61], there is no molecular symmetry and with τ values of 0.34 and 0.54, the coordination geometries follow the pattern as for **48**. The molecular packing in each of the four crystals is quite distinct. In the crystal of **47**, hydroxyl-O–H $\cdots$ O(hydroxyl) hydrogen bonding connects molecules into supramolecular layers via eight-membered { . . . H-O}$_4$ synthons. From Figure 10a, left, it is apparent that the layers have rather large voids and, indeed, perpendicularly inclined layers interpenetrate to form a three-dimensional architecture, shown in Figure 10a, right. Supramolecular layers are also formed in the R = Et analogue, **48**. The assembly is distinct, as in Figure 10b, left, from that in **47**. Centrosymmetrically related layers are connected via further hydroxyl-O–H $\cdots$ O(hydroxyl) hydrogen bonding to form double layers, shown in Figure 10b, right. The key supramolecular synthon is a 12-membered { . . . H-O}$_6$ ring connected to two further exocyclic hydroxyl-O–H $\cdots$ O(hydroxyl) hydrogen bonds. The presence of solvent methanol in **49** effectively blocks off half of the hydrogen bonding capacity of the hydroxyethyl groups owing to the formation of methanol-O–H $\cdots$ O(hydroxyl) hydrogen bonds. The remaining hydroxyethyl groups form hydroxyl-O–H $\cdots$ O(hydroxyl) hydrogen bonds to generate the supramolecular ladder shown in Figure 10c. In **50**, where both N-bound R group carry hydrogen bonding potential, a three-dimensional architecture ensues, Figure 10d. It is of interest that one of the hydroxyl groups in **50** does not form hydroxyl-O–H $\cdots$ O(hydroxyl) hydrogen bonds but hydroxyl-O–H $\cdots$ S hydrogen bonds instead. The propensity of such hydrogen bonds in $^-$S$_2$CNCH$_2$CH$_2$OH dithiocarbamate ligands has been summarised recently, a summary which suggested these are relatively common in the structural chemistry of these ligands [73].

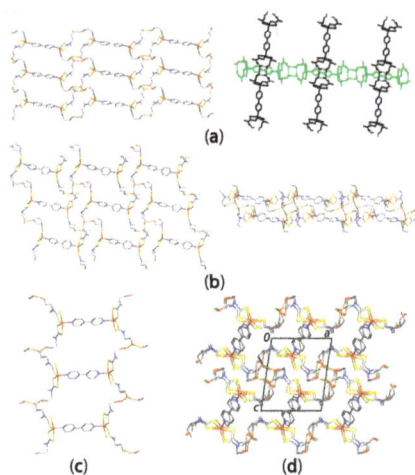

(a)

(b)

(c)   (d)

Figure 10. Supramolecular aggregation sustained by hydroxyl-O–H $\cdots$ O(hydroxyl) hydrogen bonding in the crystals of (**a**) {Zn[S$_2$CN(Me)CH$_2$CH$_2$OH]$_2$}$_2$(bpy) (**47**) showing a layer (left) and partial two-fold inclined interpenetration of two layers; (**b**) {Zn[S$_2$CN(Et)CH$_2$CH$_2$OH]$_2$}$_2$(bpy) (**48**) showing a layer (left) and the double-layer; (**c**) {Zn[S$_2$CN(Et)CH$_2$CH$_2$OH]$_2$}$_2$(bpy).2CH$_3$OH (**49**) showing a ladder; and (**d**) {Zn[S$_2$CN(CH$_2$CH$_2$OH)$_2$]$_2$}$_2$(bpy) (**50**) showing the three-dimensional architecture.

The next structure to be described was isolated from frustrated experiments to force the formation of a coordination polymer by having an excess of bpy in the crystallisation. Thus, Zn[S₂CN(iPr)CH₂CH₂OH]₂(bpy).(bpy)₀.₅ **(51)** [62] has monodentate and uncoordinated bpy in the lattice. A view of the zinc-containing molecule is shown in Figure 11a. The familiar NS₄ donor set is noted and this approaches a trigonal-bipyramidal geometry as judged by the value of τ = 0.64. There is a considerable twist in the coordinated bpy molecule of 28° which compares with 0° for the uncoordinated, centrosymmetric molecule.

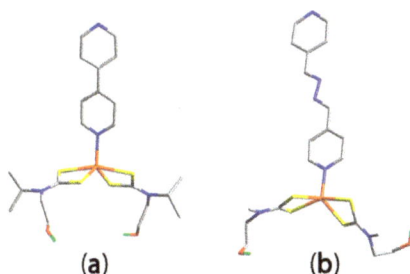

Figure 11. Aggregation in the structures of (**a**) Zn[S₂CN(iPr)CH₂CH₂OH]₂(bpy).(bpy)₀.₅ **(51)**; and (**b**) Zn[S₂CN(Me)CH₂CH₂OH]₂(4-pyald) **(52)**.

In the crystal, rather large 28-membered { ... HOC₂NCSZnSCNC₂O}₂ synthons are formed via hydroxyl-O–H ··· O(hydroxyl) hydrogen bonds as molecules arrange in a head-to-head fashion. There are two free hydroxyl-H atoms associated with this ring and they connect to the non-coordinating end of the coordinated bpy molecules via hydroxyl-O–H ··· N(pyridyl) hydrogen bonds to generate the two-dimensional array shown in Figure 12a. Weak non-covalent interactions assemble layers into a three-dimensional architecture and the connections between the host lattice and non-coordinating bpy molecules are of the type bpy-C–H···O(hydroxyl) interactions.

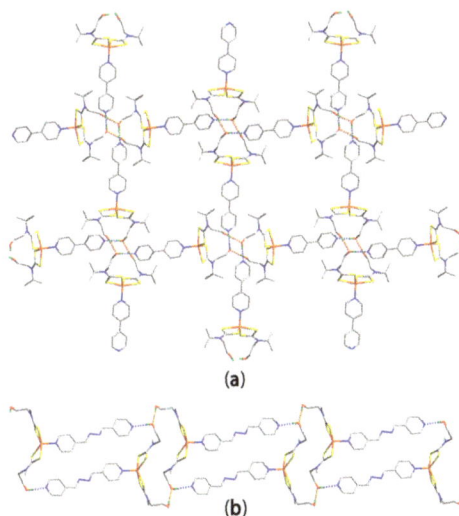

Figure 12. Supramolecular aggregation sustained by hydroxyl-O–H ··· O(hydroxyl) and hydroxyl-O–H ··· N(pyridyl) hydrogen bonding in the crystals of (**a**) Zn[S₂CN(iPr)CH₂CH₂OH]₂ (bpy).(bpy)₀.₅ **(51)**; and (**b**) Zn[S₂CN(Me)CH₂CH₂OH]₂(4-pyald) **(52)**.

There is a second structure with a monodentate bipyridyl-type molecule, namely Zn[S$_2$CN(Me)CH$_2$CH$_2$OH]$_2$(4-pyald) (**52**) [63], shown in Figure 11b. The NS$_4$ donor set defines a distorted square-pyramidal geometry (τ = 0.32). In the crystal, hydroxyl-O–H $\cdots$ O(hydroxyl) and hydroxyl-O–H $\cdots$ N(pyridyl) hydrogen bonds combine to stabilise a double-chain, Figure 12b. The 28-membered { . . . HOC$_2$NCSZnSCNC$_2$O}$_2$ synthons seen in **51** feature in the crystal of **52** also. Indeed, this synthon occurs in a fascinating pair of structures to be discussed next which combine zinc-containing molecules and bridging bipyridyl-type ligands, each with hydrogen bonding functionality.

The commonly NS$_4$ donor sets for zinc dithiocarbamates are found in the structures of Zn[S$_2$CN(Me)CH$_2$CH$_2$OH]$_2$(3-pdaO) (**53**) [64] and Zn[S$_2$CN(Me)CH$_2$CH$_2$OH]$_2$(3-pdaOt) (**54**) where 3-pdaOt, N-(pyridin-2-ylmethyl)-N$'$-(pyridin-3-ylmethyl)ethanediimidic acid, is the hydroxyl-imine tautomer of 3-pdaO, shown in Figure 13a. The values of τ for the centrosymmetric molecules in **53**, 0.66, and **54**, 0.54, suggest tendencies towards trigonal-bipyramidal. The molecular packing for both molecules is very similar and discussion will focus on that in the crystal of **54**. Here, molecules assemble via hydroxyl-O–H $\cdots$ O(hydroxyl) hydrogen bonds to form 28-membered { . . . HOC$_2$NCSZnSCNC$_2$O}$_2$ synthons and supramolecular chains, Figure 13b. In this case, the large rings are threaded by centrosymmetrically related chains to form inter-woven coordination polymers, as shown in Figure 13c. The links between the individual chains in **54** are of the type hydroxyl-O–H $\cdots$ N(imine), and in **53**, of the type hydroxyl-O–H $\cdots$ O(amide) [64]. The next two structures to be discussed contain sulphur analogues of 3-pdaO.

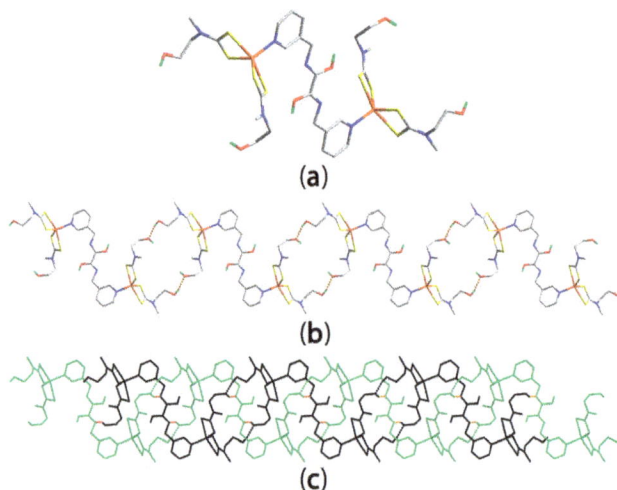

(a)

(b)

(c)

Figure 13. (**a**) Molecular structure of Zn[S$_2$CN(Me)CH$_2$CH$_2$OH]$_2$(3-pdaOt) (**54**); (**b**) supramolecular chain sustained by hydroxyl-O–H $\cdots$ O(hydroxyl) hydrogen bonding; and (**c**) schematic of the inter-woven coordination polymer with individual chains coloured black and green and with hydroxyl-O–H $\cdots$ N(imine) hydrogen bonds (orange dashed lines) between the chains shown in (**b**).

The familiar distorted NS$_4$ coordination geometries are found in each of centrosymmetric Zn[S$_2$CN(Me)CH$_2$CH$_2$OH]$_2$(3-pdaS) (**55**), Figure 14, isolated as a DMF di-solvate, and Zn[S$_2$CN(Me)CH$_2$CH$_2$OH]$_2$(3-pdaS) (**56**), which was co-crystallised serendipitously with two equivalents of S$_8$ owing to partial desulphurisation during crystallisation [65]. The values of τ are 0.46 and 0.08, respectively, suggesting the coordination geometry in **56** is relatively closely based on a square-pyramid.

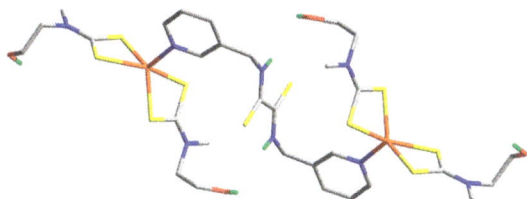

Figure 14. Molecular structure of Zn[S_2CN(Me)CH_2CH_2OH]$_2$(3-pdaS) in (**55**).

In the molecular packing of **55**, half of the hydroxyl groups form hydroxyl-O–H⋯O(DMF) hydrogen bonds and the remaining hydroxyl groups form hydroxyl-O–H⋯O(hydroxyl) hydrogen bonds to form a supramolecular ladder, via the 28-membered { ... HOC$_2$NCSZnSCNC$_2$O}$_2$ synthons commented upon above, Figure 15a. Additional stabilisation to the supramolecular one-dimensional chain is provided by amide-N–H⋯O(DMF) and amide-N–H⋯S(thioamide) hydrogen bonds, not shown. The key feature of the supramolecular aggregation in **56** is the forming of planar { ... HO}$_4$ synthons contributed to by four separate molecules and which extend laterally to form a two-dimensional array resembling that shown for **48** in Figure 10b. Layers are connected into a three-dimensional architecture by amide-N–H⋯S(amide) hydrogen bonds to define channels in which reside the S_8 molecules, Figure 15b.

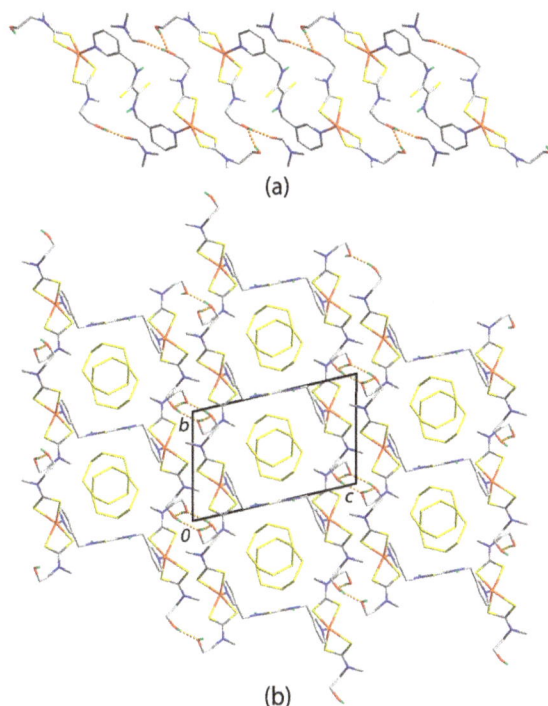

(a)

(b)

Figure 15. Supramolecular aggregation sustained by hydroxyl-O–H⋯O(hydroxyl) hydrogen bonding in the crystals of (**a**) Zn[S_2CN(Me)CH_2CH_2OH]$_2$(3-pdaS).2DMF (**55**) with additional by hydroxyl-O–H⋯O(DMF) hydrogen bonding; and (**b**) Zn[S_2CN(Me)CH_2CH_2OH]$_2$(3-pdaS).2S_8 (**56**) showing the occupancy of channels by S_8 molecules.

The final zinc-containing structure to be described is one containing the 2,4,6-tris (pyridin-2-yl)-1,3,5-triazine ligand abbreviated as triazine. The molecular structure of Zn[S$_2$CN(Me) CH$_2$CH$_2$OH]$_2$(triazine).1.5(dioxane) (**57**) [66] is noteworthy for being the only example having monodentate-coordinating dithiocarbamate ligands, as in Figure 16a. This arises as the triazine molecule is tridentate, forming bonds to zinc via a triazine- and two pyridyl-nitrogen atoms. The resulting N$_3$S$_2$ donor set is very close to a square-pyramid with τ = 0.07 and with a sulphur atom occupying the apical position. In the molecular packing, molecules are arranged in rows and are held in place by hydroxyl-O–H ··· N(triazine), N(pyridyl) hydrogen bonds, and these are connected into a two-dimensional array via hydroxyl-O–H ··· S(dithiocarbamate) hydrogen bonds as shown in Figure 16b.

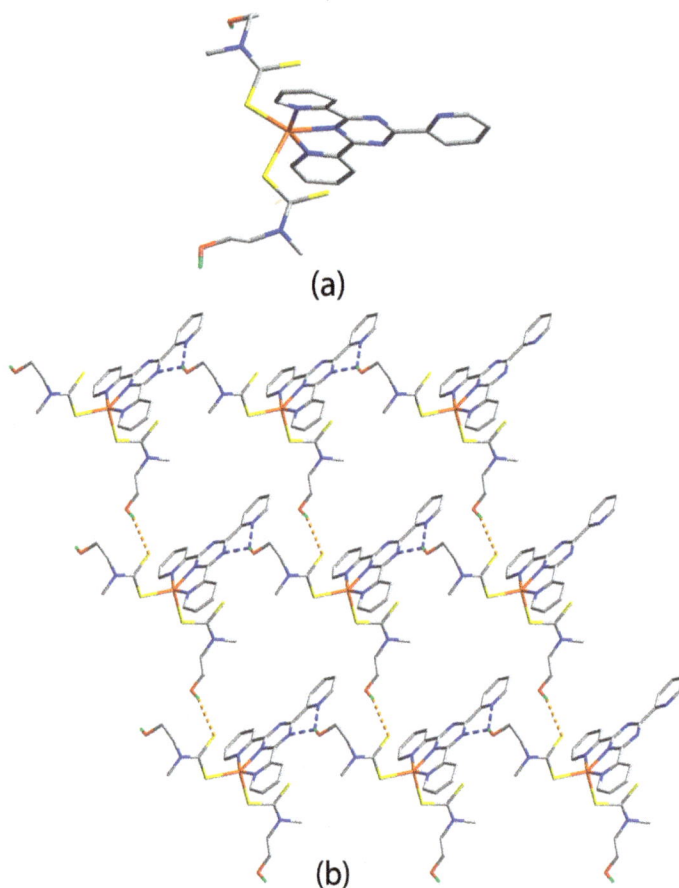

(a)

(b)

Figure 16. Images for Zn[S$_2$CN(Me)CH$_2$CH$_2$OH]$_2$(triazine)·1.5(dioxane) (**57**): (**a**) molecular structure; and (**b**) supramolecular aggregation sustained by hydroxyl-O–H ··· S(dithiocarbamate) and hydroxyl-O–H ··· S(triazine, pyridyl) hydrogen bonding shown as orange and blue dashed lines, respectively.

3.4. Cadmium Xanthate Structure

There is a sole example of a cadmium xanthate structure, [Cd(S$_2$COiPr)$_2$(bpy)]$_n$ (**58**) [74], making xanthates the least represented of the conventional 1,1-dithiolate ligands included in this review;

data for all 31 cadmium structures, that is, **58–88** [62,74–92], are listed in Table 2. There are two independent formula units in the crystal and each of these is disposed about a two-fold axis of asymmetry with the cadmium and nitrogen atoms lying on the axis. Each independent formula unit self-associates to form a linear chain. The conformations of the bridging bpy molecules distinguish the structures. In one repeat unit, the bpy is effectively planar, with the dihedral angle between the two rings being 3°, Figure 17, while, in the other residue, the rings are twisted forming a dihedral angle of 26°. This observation highlights the conformational flexibility of the bpy molecule as noted in the zinc structures described above.

Table 2. Summary of the general features of cadmium structures **58–88** [62,74–92].

Compound	R/R′	N-Ligand	Donor Set	Motif	Ref.
58	iPr	bpy	N_2S_4	linear chain	[74]
59	Me	bpy	N_2S_4	zig-zag chain	[75]
60	Et	bpy	N_2S_4	linear chain	[76]
61	iPr	bpy	N_2S_4	linear chain	[77]
62	Cy	bpy	N_2S_4	linear chain	[77]
63	iPr	bpe	N_2S_4	linear chain	[77]
64	Cy	bpe	N_2S_4	linear chain	[77]
65	iPr	bpeh	N_2S_4	linear chain	[77]
66	iPr	dpp	N_2S_4	linear chain	[77]
67 [1]	Cy	dpp	N_2S_4	dimer	[77]
68	iPr	bpS$_2$	N_2S_4	linear chain	[77]
69	iPr	4-pyald	N_2S_4	linear chain	[78]
70	iPr	3-pyald	N_2S_4	linear chain	[78]
71 [1]	Cy	3-pyald	N_2S_4	linear chain	[79]
72	iPr	2-pyald	N_2S_4	dimer	[78]
73	iPr	2-bpe	N_2S_4	linear chain	[77]
74	Cy	2-bpe	N_2S_4	linear chain	[77]
75	Me, 4-MeOPh	bpy	N_2S_4	linear chain	[80]
76	Et, 4-MeOPh	bpy	N_2S_4	linear chain	[81]
77	Me, 4-MeOPh	bpe	N_2S_4	zig-zag chain	[82]
78	Me, 4-MeOPh	dpeh	N_2S_4	zig-zag chain	[83]
79	Me, 4-MeOPh	dpp	N_2S_4	zig-zag chain	[84]
80	iBu	bpy	N_2S_4	linear chain	[85]
81	CH$_2$Ph	bpy	N_2S_4	linear chain	[86]
82	Et	bpe	N_2S_4	linear chain	[87]
83	Et	bpeh	N_2S_4	linear chain	[88]
84	nPr	2-pyald	N_2S_4	dimeric	[89]
85 [2]	iPr, CH$_2$CH$_2$OH	bpy	NS$_4$	monomeric	[62]
86 [3]	iPr, CH$_2$CH$_2$OH	bpe	N_2S_4	dimeric	[90]
87	nPr, CH$_2$CH$_2$OH	4-pyald	N_2S_4	monomeric	[91]
88 [4]	iPr, CH$_2$CH$_2$OH	3-pyald	N_3S_2	dimeric	[92]

[1] Chloroform di-solvate; [2] Co-crystallised with half an uncoordinated bpy molecule; [3] Acetonitrile tetra-solvate; [4] Dihydrate.

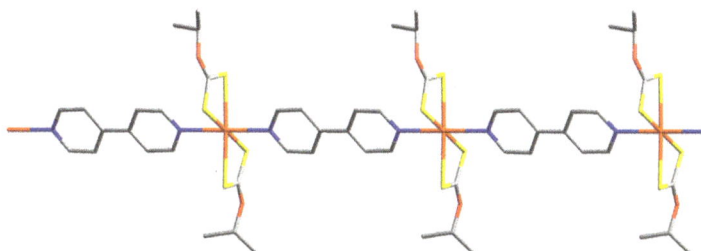

Figure 17. Aggregation in [Cd(S$_2$COiPr)$_2$(bpy)]$_n$ (**58**).

3.5. Cadmium Dithiophosphate and Related Structures

By far the most numerous cadmium-containing structures covered in this survey are the dithiophosphates, which have been the subject of a number of systematic studies. There are four structures featuring bridging bpy ligands, {Cd[S$_2$P(OR)$_2$]$_2$(bpy)}$_n$ for R = Me (**59**) [75], Et (**60**) [76], iPr (**61**) [77] and Cy (**62**) [77]. All structures present different crystallographic symmetry. Thus, in **59** the bpy ligand is disposed about a centre of inversion, in **60** the cadmium atom is located on a centre of symmetry, for **61** both bpy and cadmium are disposed about centres of inversion whereas for **62**, the cadmium and nitrogen atoms lie on a 2-fold axis of symmetry. In **59–61**, the bpy molecules are planar or effectively planar but, in **63** the dihedral angle of 50° between the rings indicates a significant twist. A common feature of the four structures is the N$_2$S$_4$ donor set for cadmium, as the dithiophosphate ligands adopt a symmetrically cheating mode, but this has a *cis*-disposition in **59**, leading to a zig-zag chain, Figure 18a, whereas *trans*-N$_2$S$_4$ geometries are found in **60–62** leading to linear topologies for the resultant coordination polymers, illustrated for **62** in Figure 18b. Linear coordination polymers and *trans*-N$_2$S$_4$ donor sets are also found in each of {Cd[S$_2$P(OiPr)$_2$]$_2$(bpe)}$_n$ (**63**) {Cd[S$_2$P(OCy)$_2$]$_2$(bpe)}$_n$ (**64**) and {Cd[S$_2$P(OiPr)$_2$]$_2$(bpeh)}$_n$ (**65**) [77]. While neither of **63** and **64** have crystallographic symmetry, each of the two independent repeat units in **65** have the cadmium atom lying on a centre of symmetry and the bpy molecule lying about a centre of inversion; independent repeat unit self-associate to form independent chains.

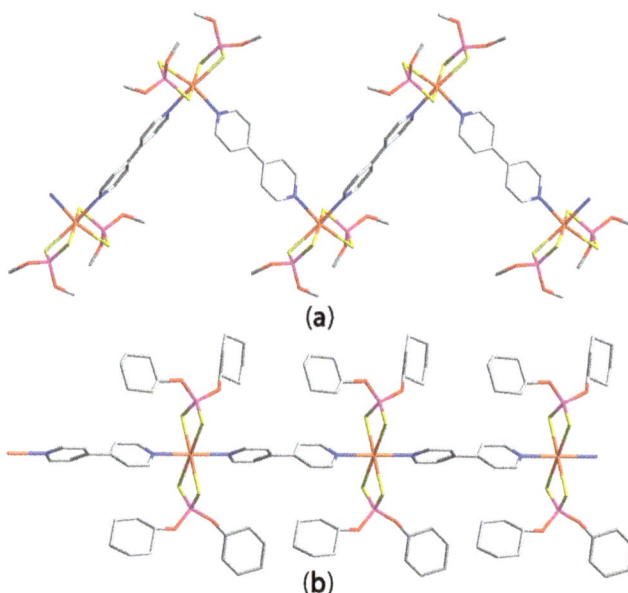

Figure 18. Aggregation in the crystals of (a) {Cd[S$_2$P(OMe)$_2$]$_2$(bpy)}$_n$ (**59**); and {Cd[S$_2$P(OCy)$_2$]$_2$(bpy)}$_n$ (**62**).

There are two structures available with the dpp ligand and the supramolecular aggregation in these are dictated by steric effects. In {Cd[S$_2$P(OiPr)$_2$]$_2$(bpp)}$_n$ (**66**) [77], with the cadmium atom lying on a 2-fold axis of symmetry and with the dpp bisected by a 2-fold axis of symmetry, a *cis*-N$_2$S$_4$ donor set is found resulting in a linear coordination polymer with characteristic arches owing to the conformation of the dpp ligand, Figure 19a. A similar *cis*-N$_2$S$_4$ donor set and arched conformation is found in the structure of {Cd[S$_2$P(OCy)$_2$]$_2$(bpp)}$_2$ (**67**) [77], but not the formation of a coordination polymer owing to the steric bulk of the cyclohexyl groups. Instead, a centrosymmetric dimeric aggregate as shown in Figure 19b. In {Cd[S$_2$P(OiPr)$_2$]$_2$(bpS$_2$)}$_n$ (**68**) [77], which is devoid of crystallographic symmetry, a linear coordination polymer is formed. As seen in Figure 19c, this topology resembles that found for **66** illustrated in Figure 19a. The structures of {Cd[S$_2$P(OiPr)$_2$]$_2$(4-pyald)}$_n$ (**69**) and the 3-pyald analogue (**70**) [78] share common features in terms of symmetry (the cadmium atom lies on a centre of symmetry and the 4-pyald molecule lies about a centre of inversion), *trans*-N$_2$S$_4$ coordination geometry which defines an octahedron and linear topology for the resultant coordination polymer compared with Figure 18b. A similar description pertains for {Cd[S$_2$P(OCy)$_2$]$_2$(4-pyald)}$_n$ (**71**) [79] but, in this case only the cadmium atom lies on a centre of inversion. A different motif is formed when reactions are conducted with the isomeric 2-pyald ligand. As shown in Figure 19d, all four nitrogen atoms of 2-pyald are employed in coordination to cadmium, which exists within a *cis*-N$_2$S$_4$ donor set in the centrosymmetric dimer, {Cd[S$_2$P(OiPr)$_2$]$_2$(2-pyald)}$_2$ (**72**) [78]. The final two structures are also 2-pyridyl isomers of the more commonly explored bpe ligand. In both {Cd[S$_2$P(OR)$_2$]$_2$(2-bpe)}$_n$ for R = iPr (**73**) [79], Figure 19e, and Cy (**74**) [77], the cadmium (within *trans*-N$_2$S$_4$ donor sets) and 2-bpe molecules are disposed on/about centres of inversion. Linear coordination polymers are formed as the chelation observed in **72** is no longer possible, thereby making available additional coordination sites for bridging.

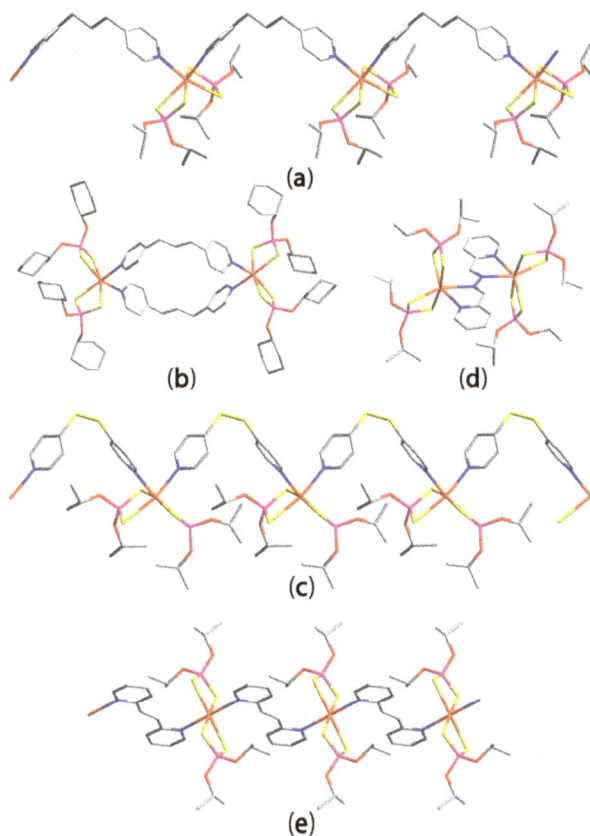

Figure 19. Aggregation in the crystals of (**a**) {Cd[S$_2$P(OiPr)$_2$]$_2$(bpp)}$_n$ (**66**); (**b**) {Cd[S$_2$P(OCy)$_2$]$_2$(bpp)}$_2$ (**67**); (**c**) {Cd[S$_2$P(OiPr)$_2$]$_2$(bpS$_2$)}$_n$ (**68**); (**d**) {Cd[S$_2$P(OiPr)$_2$]$_2$(2-pyald)}$_2$ (**72**); and (**e**) Cd[S$_2$P(OiPr)$_2$]$_2$ (2-bpe)}$_n$ (**73**).

There are five dithiophosphonate structures and one example of a dithiophosphinate structure available in the CSD [2] and these uniformly are one-dimensional coordination polymers but, with distinct topologies. The structures of two bpy-containing species {Cd[S$_2$P(OR)4-MeOPh]$_2$(bpy)}$_n$ for R = Me (**75**) [80] and Et (**76**) [81] are linear polymers, each with the *trans*-N$_2$S$_4$ octahedral cadmium atom located on a centre of inversion and bpy situated about a centre. In the same way, the three structures of general formula {Cd[S$_2$P(OMe)4-MeOPh]$_2$(N∩N)}$_n$ for N∩N = bpe (**77**) [82], beph (**78**) [83] and dpp (**79**) [84] are very similar, all lacking symmetry, adopting zig-zag polymers and each with *cis*-N$_2$S$_4$ octahedral cadmium. The sole example of a dithiophosphinate, namely {Cd[S$_2$P(OiBu)$_2$]$_2$(bpy)}$_n$ (**80**) [85], is distinct from the aforementioned structures in that the CdNNCd axis lies on a two-fold axis of symmetry implying the coordination polymer is linear; the cadmium atom is octahedrally coordinated within a *trans*-N$_2$S$_4$ donor set.

3.6. Cadmium Dithiocarbamate Structures

As for the zinc dithiocarbamates, this section is divided into a discussion of structures incapable of hydrogen bonding and those that are capable of hydrogen bonding. In the latter category, a number of unprecedented motifs are apparent.

3.6.1. Cadmium Dithiocarbamate Structures Not Capable of Forming Hydrogen Bonds

There are only four structures in this category with four different bipyridyl-type ligands. The common feature of {Cd[S$_2$CN(CH$_2$Ph)$_2$]$_2$(bpy)}$_n$ (**81**) [86], with the CdNNCd axis lying on a 2-fold axis of symmetry, [Cd(S$_2$CNEt$_2$)$_2$(bpe)]$_n$ (**82**) [87], with two half, centrosymmetric bpe molecules in the asymmetric unit along with Cd(S$_2$CNEt$_2$)$_2$, and [Cd(S$_2$CNEt$_2$)$_2$(bpeh)]$_n$ (**83**) [88], with two independent repeat units in the asymmetric unit, Figure 20a, is the presence of *trans*-N$_2$S$_4$ cadmium centres and linear coordination polymers. The fourth structure, {Cd[S$_2$CN(nPr)$_2$]$_2$(2-pyald)}$_2$ (**84**) [83], Figure 20b, closely resembles that of {Cd[S$_2$P(OiPr)$_2$]$_2$(2-pyald)}$_2$ (**72**) [78], being a centrosymmetric dimer with a *cis*-N$_2$S$_4$ donor set owing to chelating 2-pyald ligands.

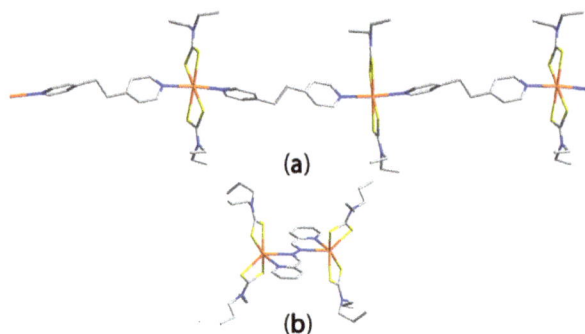

Figure 20. Supramolecular aggregation in the crystals of (a) [Cd(S$_2$CNEt$_2$)$_2$(bpeh)]$_n$ (**83**); and (b) {Cd[S$_2$CN(nPr)$_2$]$_2$(2-pyald)}$_2$ (**84**).

3.6.2. Cadmium Dithiocarbamate Structures Capable of Forming Hydrogen Bonds

The final four structures to be described in this bibliographic review are zero-dimensional and are perhaps not of great interest in terms of the generation of coordination polymers. However, the isolation of these species raises questions relating to the relative importance of coordinate-bond formation versus hydrogen bonding interactions and, therefore, these are of fundamental interest. The molecular structures of {Cd[S$_2$CN(iPr)CH$_2$CH$_2$OH]$_2$(bpy)}$_2$ (**85**) [69], co-crystallised with half a bpy molecule, {Cd[S$_2$CN(iPr)CH$_2$CH$_2$OH]$_2$}$_2$(bpe)$_3$ (**86**) [90], {Cd[S$_2$CN(nPr)CH$_2$CH$_2$OH]$_2$}(4-pyald)$_2$ (**87**) [91] and {Cd[S$_2$CN(iPr)CH$_2$CH$_2$OH]$_2$}$_2$(3-pyald)$_2$ (**88**) [92] are illustrated in Figure 21a–d. In **85**, which features a monodentate bpy ligand, the cadmium atom is five-coordinate within a NS$_4$ donor set that defines a geometry approaching trigonal-bipyramidal ($\tau = 0.67$) with the less tightly bound sulphur atoms defining the axial positions. A distorted *cis*-N$_2$S$_4$ donor set is found in centrosymmetric and binuclear **86** which features a bridging bpe ligand and two terminally bound bpe molecules. An octahedral geometry is also found in **87** but, based on a *trans*-N$_2$S$_4$ donor set as the cadmium centre, located on a centre on inversion, is coordinated by two monodentate 4-pyald molecules. Finally, and quite remarkably, the structure of centrosymmetric **88**, features octahedral cadmium centres within NS$_5$ donor sets as the 3-pyald molecules are coordinating in the monodentate mode. In all of the previous **87** structures described in this review, the original parent structure, which is often aggregated [25], has been disrupted by the addition of base. In **88**, the thermodynamically favoured dimeric structure, as opposed to the initially formed one-dimensional coordination polymer [93,94], is retained but, with the addition of base in the form of 3-pyald to increase the normally observed S$_5$ donor set to NS$_5$. These four structures, again, point to the unpredictability of this chemistry and substantiate the need of systematic studies. It is probably of no coincidence that the aforementioned four structures each feature hydrogen bonding capability in the dithiocarbamate ligands.

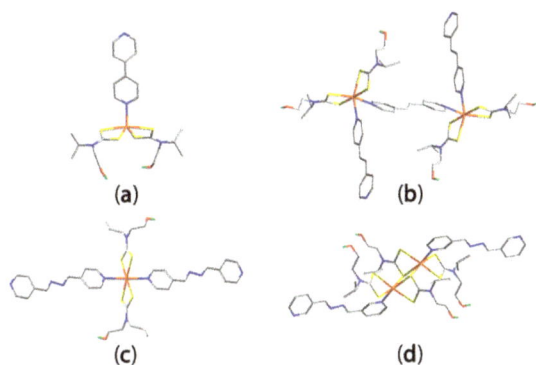

Figure 21. Aggregation in the crystals of (a) {Cd[S$_2$CN(iPr)CH$_2$CH$_2$OH]$_2$(bpy)}$_2$ (**85**); (b) {Cd[S$_2$CN(iPr)CH$_2$CH$_2$OH]$_2$}$_2$(bpe)$_3$ (**86**); (c) {Cd[S$_2$CN(nPr)CH$_2$CH$_2$OH]$_2$}(4-pyald)$_2$ (**87**); and (d) {Cd[S$_2$CN(iPr)CH$_2$CH$_2$OH]$_2$}$_2$(3-pyald)$_2$ (**88**).

To date, not much mention of the synthesis conditions has been made. Here, it is pertinent to note that the stoichiometry of the reactants in these reactions with functionalised dithiocarbamate ligands does not necessarily determine the synthetic outcome. For example, crystals of **86** can be isolated from solutions containing 2:1, 1:1 and 1:2 ratios of "Cd[S$_2$N(iPr)CH$_2$CH$_2$OH]$_2$" and bpe [90].

The structure of **85** is isostructural with the zinc analogue, **51**, and therefore the description of the molecular packing is identical to that described above, see Figure 12a. In the crystal of **86**, supramolecular layers are formed mediated by hydroxyl-O–H $\cdots$ O(hydroxyl) and hydroxyl-O–H $\cdots$ N(bpe) hydrogen bonding; that is, with the nitrogen acceptors derived from the monodentate bpe ligands, shown in Figure 22a. Layers interpenetrate to yield a three-dimensional structure, as shown in Figure 22b.

In the crystal of **87**, only hydroxyl-O–H $\cdots$ N(4-pyald) hydrogen bonds are formed and these lead to 40-membered [CdSCNC$_2$OH $\cdots$ NC$_4$N$_2$C$_4$N]$_2$ synthons and supramolecular ladders as shown in Figure 23a. In the crystal of **88**, supramolecular layers sustained by hydroxyl-O–H $\cdots$ O(hydroxyl) and water-O–H $\cdots$ O(hydroxyl) hydrogen bonding are connected into a three-dimensional architecture via water-O–H $\cdots$ N(pyridyl) hydrogen bonding, as in Figure 23b.

(a)

(b)

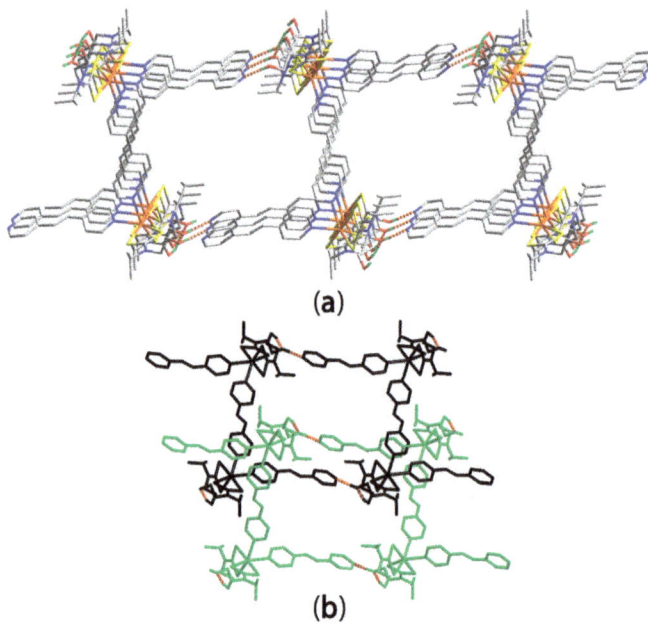

Figure 22. Supramolecular aggregation in the crystals of {Cd[S$_2$CN(iPr)CH$_2$CH$_2$OH]$_2$}$_2$(bpe)$_3$ (**86**) (**a**) layers sustained by hydroxyl-O–H $\cdots$ O(hydroxyl) and hydroxyl-O–H $\cdots$ N(bpe) hydrogen bonding; and (**b**) two-fold interpenetration of two supramolecular layers.

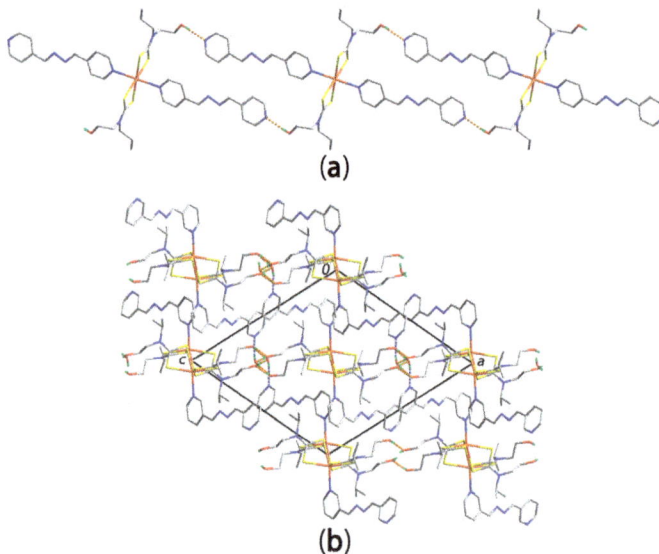

(a)

(b)

Figure 23. Supramolecular aggregation in the crystals of (**a**) {Cd[S$_2$CN(nPr)CH$_2$CH$_2$OH]$_2$}(4-pyald)$_2$ (**87**); and (**b**) {Cd[S$_2$CN(iPr)CH$_2$CH$_2$OH]$_2$}$_2$(3-pyald)$_2$ (**88**).

4. Overview and Conclusions

The foregoing overview of the structural chemistry of zinc and cadmium 1,1-dithiolates with potentially bridging bipyridyl-type ligands reveals a very real dependence upon aggregation patterns on the nature of the 1,1-dithiolate ligand as well as upon the metal centre. The maximum dimension for any coordination polymer in the structures is one, and for zinc, polymers are only formed for xanthates and dithiophosphates. For zinc dithiocarbamates, only zero-dimensional aggregates are formed; an observation which is readily rationalised in terms of the relatively greater coordination potential of the dithiocarbamate ligand compared with the other 1,1-dithiolates. This greater chelating ability arises owing to the relative importance of the resonance form where there are two formal negative charges on the sulphur atoms, Figure 24. The situation changes for cadmium where one-dimensional coordination polymers are found for all 1,1-dithiolate ligands, an observation correlating with the larger size of cadmium versus zinc.

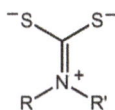

Figure 24. Chemical diagram of a significant resonance structure for the dithiocarbamate ligand. R, R′ = alkyl, aryl.

A wide array of coordination geometries are observed across the series, often arising from flexible coordination modes of xanthate and dithiophosphate ligands which range from monodentate to chelating and sometimes involve the alkoxy-O atoms in coordination. The steric role of 1,1-dithiolate-bound R groups was shown to be instrumental in controlling aggregation patterns in that larger groups, such as cyclohexyl, preclude the close approach of entities and thereby, exert an influence both over aggregation and polymer topology. A fascinating observation is made for dithiocarbamate ligands functionalised with hydrogen bonding potential, in that an apparent competition between the formation of M←N(pyridyl) coordinate-bonds and hydroxyl-O–H ⋯ N(pyridyl) hydrogen bonds exists leading to monodentate binding modes for the bipyridyl-type ligands. Such a competition might indicate the energy of stabilisation of these bonding modes are similar. Certainly, the unexpected structures demand further systematic studies as more insight will be gained into the fundamental coordination chemistry of metal 1,1-dithiolates.

Acknowledgments: Sunway University is gratefully acknowledged for continuing support for chemical crystallography studies.

Conflicts of Interest: The author declares no conflict of interest.

References

1. Moghadam, P.Z.; Li, A.; Wiggin, S.B.; Tao, A.; Maloney, A.G.P.; Wood, P.A.; Ward, S.C.; Fairen-Jimenez, D. Development of a Cambridge Structural Database subset: A collection of metal-organic frameworks for past, present, and future. *Chem. Mater.* **2017**, *29*, 2618–2625. [CrossRef]
2. Groom, C.R.; Bruno, I.J.; Lightfoot, M.P.; Ward, S.C. The Cambridge Structural Database. *Acta Crystallogr. B Struct. Sci. Cryst. Eng. Mater.* **2016**, *72*, 171–179. [CrossRef] [PubMed]
3. Wang, L.; Han, Y.Z.; Feng, X.; Zhou, J.W.; Qi, P.F.; Wang, B. Metal-organic frameworks for energy storage: Batteries and supercapacitors. *Coord. Chem. Rev.* **2016**, *307*, 361–381. [CrossRef]
4. Kumar, P.; Kim, K.-H.; Kwon, E.E.; Szulejko, J.E. Metal-organic frameworks for the control and management of air quality: Advances and future direction. *J. Mater. Chem. A* **2016**, *4*, 345–361. [CrossRef]
5. Noh, T.H.; Jung, O.S. Recent advances in various metal-organic channels for photochemistry beyond confined spaces. *Acc. Chem. Res.* **2016**, *49*, 1835–1843. [CrossRef] [PubMed]

6. Rogge, S.M.J.; Bavykina, A.; Hajek, J.; Garcia, H.; Olivos-Suarez, A.I.; Sepulveda-Escribano, A.; Vimont, A.; Clet, G.; Bazin, P.; Kapteijn, F. Metal-organic and covalent organic frameworks as single-site catalysts. *Chem. Soc. Rev.* **2017**, *46*, 3134–3184. [CrossRef] [PubMed]

7. D'Vries, R.F.; Iglesias, M.; Snejko, N.; Gutiérrez-Puebla, E.; Monge, M.A. Lanthanide metal-organic frameworks: Searching for efficient solvent-free catalysts. *Inorg. Chem.* **2012**, *51*, 11349–11355. [CrossRef] [PubMed]

8. He, C.; Liu, D.; Lin, W. Nanomedicine applications of hybrid nanomaterials built from metal-ligand coordination bonds: Nanoscale metal-organic frameworks and nanoscale coordination polymers. *Chem. Rev.* **2015**, *115*, 11079–11108. [CrossRef] [PubMed]

9. Johnstone, T.C.; Suntharalingam, K.; Lippard, S.J. The next generation of platinum drugs: Targeted Pt(II) agents, nanoparticle delivery, and Pt(IV) prodrugs. *Chem. Rev.* **2016**, *116*, 3436–3486. [CrossRef] [PubMed]

10. Wyszogrodzka, G.; Marszałek, B.; Gil, B.; Dorożyński, P. Metal-organic frameworks: Mechanisms of antibacterial action and potential applications. *Drug Discov. Today* **2016**, *21*, 1009–1018. [CrossRef] [PubMed]

11. Wang, H.-S. Metal-organic frameworks for biosensing and bioimaging applications. *Coord. Chem. Rev.* **2017**, *349*, 139–155. [CrossRef]

12. Anderson, S.L.; Stylianou, K.C. Biologically derived metal organic frameworks. *Coord. Chem. Rev.* **2017**, *349*, 102–128. [CrossRef]

13. Lian, X.; Fang, Y.; Joseph, E.; Wang, Q.; Li, J.L.; Banerjee, S.; Lollar, C.; Wang, X.; Zhou, H.-C. Enzyme-MOF (metal-organic framework) composites. *Chem. Soc. Rev.* **2017**, *46*, 3386–3401. [CrossRef] [PubMed]

14. Inokuma, Y.; Arai, T.; Fujita, M. Networked molecular cages as crystalline sponges for fullerenes and other guests. *Nat. Chem.* **2010**, *2*, 780–783. [CrossRef] [PubMed]

15. Hoshino, M.; Khutia, A.; Xing, H.Z.; Inokuma, Y.; Fujita, M. The crystalline sponge method updated. *IUCrJ* **2016**, *3*, 139–151. [CrossRef] [PubMed]

16. Eddaoudi, M.; Moler, D.B.; Li, H.; Chen, B.; Reineke, T.M.; O'Keeffe, M.; Yaghi, O.M. Modular chemistry: Secondary building units as a basis for the design of highly porous and robust metal-organic carboxylate frameworks. *Acc. Chem. Res.* **2001**, *34*, 319–330. [CrossRef] [PubMed]

17. Furukawa, H.; Cordova, K.E.; O'Keeffe, M.; Yaghi, O.M. The chemistry and applications of metal-organic frameworks. *Science* **2013**, *341*, 6149. [CrossRef] [PubMed]

18. Moulton, B.; Zaworotko, M.J. From molecules to crystal engineering: Supramolecular isomerism and polymorphism in network solids. *Chem. Rev.* **2001**, *101*, 1629–1658. [CrossRef] [PubMed]

19. Perry, J.J.; Perman, J.A.; Zaworotko, M.J. Design and synthesis of metal-organic frameworks using metal-organic polyhedra as supermolecular building blocks. *Chem. Soc. Rev.* **2009**, *38*, 1400–1417. [CrossRef] [PubMed]

20. Tiekink, E.R.T.; Haiduc, I. Stereochemical aspects of metal xanthate complexes: Molecular structures and supramolecular self-assembly. *Prog. Inorg. Chem.* **2005**, *54*, 127–319. [CrossRef]

21. Haiduc, I.; Sowerby, D.B. Stereochemical aspects of phosphor-1,1-dithiolato metal complexes: Coordination patterns, molecular structures and supramolecular associations in dithiophosphinates and related compounds. *Polyhedron* **1996**, *15*, 2469–2521. [CrossRef]

22. Van Zyl, W.E.; Woollins, J.D. The coordination chemistry of dithiophosphonates: An emerging and versatile ligand class. *Coord. Chem. Rev.* **2013**, *257*, 718–731. [CrossRef]

23. Heard, P.J. Main Group Dithiocarbamate Complexes. *Prog. Inorg. Chem.* **2005**, *53*, 1–69. [CrossRef]

24. Hogarth, G. Transition Metal Dithiocarbamates: 1978–2003. *Prog. Inorg. Chem.* **2005**, *53*, 71–561. [CrossRef]

25. Tiekink, E.R.T. Molecular architecture and supramolecular association in the zinc-triad 1,1-dithiolates. Steric control as a design element in crystal engineering? *CrystEngComm* **2003**, *5*, 101–113. [CrossRef]

26. Spek, A.L. Structure validation in chemical crystallography. *Acta Crystallogr. D Biol. Crystallogr.* **2009**, *65*, 148–155. [CrossRef] [PubMed]

27. *DIAMOND*, version 2.1b; K. Brandenburg & M. Berndt GbR: Bonn, Germany, 2006.

28. Ara, I.; El Bahij, F.; Lachkar, M. Synthesis, characterization and X-ray crystal structures of new ethylxanthato complexes of zinc(II) with N-donor ligands. *Synth. React. Inorg. Met.-Org. Nano-Met. Chem.* **2006**, *36*, 399–406. [CrossRef]

29. Klevtsova, R.F.; Leonova, T.G.; Glinskaya, L.A.; Larionov, S.V. Synthesis of heteroligand complexes of zinc(II) alkylxanthates with 4,4'-bipyridine. The crystal and molecular structure of the $Zn_2(4,4'$-Bipy) (i-$C_3H_7OCS_2)_4 \cdot CH_2Cl_2$ solvate. *Russ. J. Coord. Chem.* **2000**, *26*, 172–177.

30. Larionov, S.V.; Glinskaya, L.A.; Leonova, T.G.; Klevtsova, R.F. Polymeric structure of the complex [Zn(4,4′-Bipy)(i-BuOCS$_2$)$_2$]$_n$ with monodentate isobutylxanthate ligands. *Russ. J. Coord. Chem.* **1998**, *24*, 851–856.

31. Kang, J.-G.; Shin, J.-S.; Cho, D.-H.; Jeong, Y.-K.; Park, C.; Soh, S.F.; Lai, C.S.; Tiekink, E.R.T. Steric control over supramolecular polymer formation in trans-1,2-bis(4-pyridyl)ethylene adducts of zinc xanthates: Implications for luminescence. *Cryst. Growth Des.* **2010**, *10*, 1247–1256. [CrossRef]

32. Chen, D.; Lai, C.S.; Tiekink, E.R.T. Supramolecular aggregation in diimine adducts of zinc(II) dithiophosphates: Controlling the formation of monomeric, dimeric, polymeric (zig-zag and helical), and 2-D motifs. *CrystEngComm* **2006**, *8*, 51–58. [CrossRef]

33. Zhu, D.-L.; Yu, Y.-P.; Guo, G.-C.; Zhuang, H.-H.; Huang, J.-S.; Liu, Q.; Xu, Z.; You, X.-Z. *catena*-Poly[bis(*O,O′*-diethyldithiophosphato-*S*)zinc(II)-μ-4,4′-bipyridyl-*N:N′*]. *Acta Crystallogr. C Struct. Chem.* **1996**, *52*, 1963–1966. [CrossRef]

34. Glinskaya, L.A.; Shchukin, V.G.; Klevtsova, R.F.; Mazhara, A.N.; Larionov, S.V. Synthesis and polymer structure of [Zn(4,4′-bipy){(i-PrO)$_2$PS$_2$}$_2$]$_n$ and thermal properties of ZnL{(i-PrO)$_2$PS$_2$}$_2$ (L = phen, 2,2′-bipy, 4,4′-bipy). *J. Struct. Chem.* **2000**, *41*, 632–639. [CrossRef]

35. Lai, C.S.; Liu, S.; Tiekink, E.R.T. Steric control over polymer formation and topology in adducts of zinc dithiophosphates formed with bridging bipyridine ligands. *CrystEngComm* **2004**, *6*, 221–226. [CrossRef]

36. Welte, W.B.; Tiekink, E.R.T. *catena*-Poly[[bis(*O,O′*-diisopropyl dithiophosphato-κ^2*S,S′*)zinc(II)]-μ-1,2-di-4-pyridylethylene-κ^2*N:N′*]. *Acta Crystallogr. E Crystallogr. Commun.* **2007**, *63*, m790–m792. [CrossRef]

37. Welte, W.B.; Tiekink, E.R.T. *catena*-Poly[[bis(*O,O′*-diisobutyl dithiophosphato-κ^2*S,S′*)zinc(II)]-μ-1,2-di-4-pyridylethylene-κ^2*N:N′*]. *Acta Crystallogr. E Crystallogr. Commun.* **2006**, *62*, m2070–m2072. [CrossRef]

38. Tiekink, E.R.T.; Wardell, J.L.; Welte, W.B. *catena*-Poly[[bis(*O,O′*-diethyl dithiophosphato-κ^2*S,S′*)zinc(II)]-μ-1,2-di-4-pyridylethane-κ^2*N:N′*]. *Acta Crystallogr. E Crystallogr. Commun.* **2007**, *63*, m818–m820. [CrossRef]

39. Lai, C.S.; Liu, S.; Tiekink, E.R.T. *catena*-Poly[[bis(*O,O′*-diisobutyldithiophosphato-κ^2*S,S′*)zinc(II)]-μ-1,2-bis(4-pyridyl)ethane-κ^2*N:N′*]. *Acta Crystallogr. E Crystallogr. Commun.* **2004**, *60*, m1005–m1007. [CrossRef]

40. Chen, D.; Lai, C.S.; Tiekink, E.R.T. *catena*-Poly[[bis(*O,O′*-dicyclohexyldithiophosphato-κ^2*S,S′*)zinc(II)]-μ-1,2-bis(4-pyridylmethylene)hydrazine-κ^2*N:N′*]. *Acta Crystallogr. E Crystallogr. Commun.* **2005**, *61*, m2052–m2054. [CrossRef]

41. Avila, V.; Tiekink, E.R.T. *catena*-Poly[[bis(*O,O′*-diisopropyl dithiophosphato-κ^2*S,S′*)zinc(II)]-μ-1,2-bis(3-pyridylmethylene)hydrazine-κ^2*N:N′*]. *Acta Crystallogr. E Crystallogr. Commun.* **2006**, *62*, m3530–m3531. [CrossRef]

42. Devillanova, F.; Aragoni, C.; Arca, M.; Huth, S.L.; Hursthouse, M.B. CAGLIARI 154a - C$_{28}$H$_{32}$N$_2$O$_4$P$_2$S$_4$Zn$_1$. *Pers. Commun. Camb. Str. Database* **2008**. [CrossRef]

43. Shchukin, V.G.; Glinskaya, L.A.; Klevtsova, R.F.; Larionov, S.V. Synthesis, structure, and thermal properties of heteroligand coordination compounds ZnL{(i-C$_4$H$_9$)$_2$PS$_2$}$_2$ (L = Phen, 2,2′-Bipy, or 4,4′-Bipy). Polymeric structure of [Zn(4,4′-Bipy){(i-C$_4$H$_9$)$_2$PS$_2$}$_2$]$_n$. *Russ. J. Coord. Chem.* **2000**, *26*, 331–337.

44. Klevtsova, R.F.; Glinskaya, L.A.; Berus, E.I.; Larionov, S.V. Monodentate and bridging bidentate functions of 4,4′-bipyridine in the crystal structures of [Zn(4,4′-Bipy){(n-C$_3$H$_7$)$_2$NCS$_2$}$_2$] and [Zn$_2$(4,4′-Bipy){(n-C$_3$H$_7$)$_2$NCS$_2$}$_4$]. *J. Str. Chem.* **2001**, *42*, 639–647. [CrossRef]

45. Zha, M.-Q.; Li, X.; Bing, Y.; Lu, Y. μ-4,4′-Bipyridine-κ^2*N:N′*)bis[bis(*N,N*-dimethyldithiocarbamato-κ^2*S,S′*)zinc(II)]. *Acta Crystallogr. E Crystallogr. Commun.* **2010**, *66*, m1465. [CrossRef]

46. Zemskova, S.M.; Glinskaya, L.A.; Durasov, V.B.; Klevtsova, R.F.; Larionov, S.V. Mixed-ligand complexes of zinc(II) and cadmium(II) diethyldithiocarbamates with 2,2′-bipyridyl and 4,4′-bipyridyl-synthesis, structure, and thermal-properties. *J. Struct. Chem.* **1993**, *34*, 794–802. [CrossRef]

47. Lai, C.S.; Tiekink, E.R.T. (4,4′-Bipyridine)bis[bis(*N,N*-diethyldithiocarbamato)zinc(II)]. *Appl. Organomet. Chem.* **2003**, *17*, 253–254. [CrossRef]

48. Larionov, S.V.; Klevtsova, R.F.; Shchukin, V.G.; Glinskaya, L.A.; Zemskova, S.M. Heteroligand complexes of Zn(II) diisopropyl dithiocarbamate with 1,10-phenanthroline and 4,4′-bipyridlne. Crystal and molecular structures of clathrate compound Zn(4,4′-Bipy)((i-C$_3$H$_7$)$_2$NCS$_2$)$_4$·2C$_6$H$_5$CH$_3$. *Russ. J. Coord. Chem.* **1999**, *25*, 694–700.

49. Zemskova, S.M.; Glinskaya, L.A.; Klevtsova, R.F.; Durasov, V.B.; Gromilov, S.A.; Larionov, S.V. Volatile mixed-ligand complexes of bis(diisobutyldithiocarbamato)zinc with 1,10-phenanthroline, 2,2'-bipyridyl, and 4,4'-bipyridyl. Crystal and molecular structure of the binuclear complex [$Zn_2(C_{10}H_8N_2)\{(i\text{-}C_4H_9)2NCS_2\}_4$]. *J. Struct. Chem.* **1996**, *37*, 941–947. [CrossRef]

50. Yin, X.; Zhang, W.; Fan, J.; Wei, F.X.; Lai, C.S.; Tiekink, E.R.T. Bis[bis(*N,N*-dibenzyldithiocarbamato)zinc(II)] (4,4'-bipyridine). *Appl. Organomet. Chem.* **2003**, *17*, 889–890. [CrossRef]

51. Chen, X.-F.; Liu, S.-H.; Zhu, X.-H.; Vittal, J.J.; Tan, G.-K.; You, X.-Z. μ-(4,4'-Bipyridine)-*N*:*N'*-bis [bis-(pyrrolidinedithiocarboxylato-*S*,*S'*)zinc(II)]. *Acta Crystallogr. C Struct. Chem.* **2000**, *56*, 42–43. [CrossRef]

52. Liu, S.-H.; Chen, X.-F.; Zhu, X.-H.; Duan, C.-Y.; You, X.-Z. Crystal structure and thermal analysis of a 4,4'-bipy-bridged binuclear zinc(II) complex, 2[R_2NCS_2]$_2$·Zn(4,4'-bipy) (R = piperidyl). *J. Coord. Chem.* **2001**, *53*, 223–231. [CrossRef]

53. Poplaukhin, P.; Tiekink, E.R.T. (μ-1,2-Di-4-pyridylethylene-κ^2N:*N'*)bis[bis(*N,N*-dimethyldithiocarbamato-κ^2S,*S'*)zinc(II)]. *Acta Crystallogr. E Crystallogr. Commun.* **2009**, *65*, m1474. [CrossRef]

54. Arman, H.D.; Poplaukhin, P.; Tiekink, E.R.T. (μ-trans-1,2-Di-4-pyridylethylene-κ^2N:*N'*)bis[bis(*N,N*-diethyldithiocarbamato-κ^2S,*S'*)zinc(II)] chloroform solvate. *Acta Crystallogr. E Crystallogr. Commun.* **2009**, *65*, m1472–m1473. [CrossRef]

55. Arman, H.D.; Poplaukhin, P.; Tiekink, E.R.T. (μ-trans-1,2-Di-4-pyridylethylene-κ^2N:*N'*)bis[bis(*N,N*-diisopropyldithiocarbamato-κ^2S,*S'*)zinc(II)]. *Acta Crystallogr. E Crystallogr. Commun.* **2009**, *65*, m1475. [CrossRef]

56. Lai, C.S.; Tiekink, E.R.T. Bis[bis(*N,N*-diethyldithiocarbamato)zinc(II)] (trans-1,2-bis(4-pyridyl)ethylene) trans-1,2-bis(4-pyridyl)ethylene lattice adduct. *Appl. Organomet. Chem.* **2003**, *17*, 251–252. [CrossRef]

57. Avila, V.; Tiekink, E.R.T. μ-1,2-Di-4-pyridylethane-κ^2N:*N'*-bis[bis(*N,N*-diisopropyldithiocarbamato-κ^2S,*S'*) zinc(II)]. *Acta Crystallogr. E Crystallogr. Commun.* **2008**, *64*, m680. [CrossRef]

58. Arman, H.D.; Poplaukhin, P.; Tiekink, E.R.T. Crystal structure of (μ$_2$-*N,N'*-bis((pyridin-4-yl)methyl) ethanediamide-κ^2N:*N'*)-tetrakis(diethylcarbamodithioato-κ^2S,*S'*)dizinc(II), $C_{34}H_{54}N_8O_2S_8Zn_2$. *Z. Krist. New Cryst. Struct.* **2017**, in press. [CrossRef]

59. Arman, H.D.; Poplaukhin, P.; Tiekink, E.R.T. Crystal structures of {μ$_2$-*N,N'*-bis[(pyridin-3-yl)-methyl] ethanediamide}tetrakis(dimethylcarbamodithioato)dizinc(II) dimethylformamide disolvate and {{μ$_2$-*N,N'*-bis[(pyridin-3-yl)methyl]ethanediamide}tetrakis(di-*n*-propylcarbamodithioato)-dizinc(II). *Acta Crystallogr. E Crystallogr. Commun.* **2017**, *73*, 1501–1507. [CrossRef] [PubMed]

60. Jotani, M.M.; Poplaukhin, P.; Arman, H.D.; Tiekink, E.R.T. Supramolecular association in (μ$_2$-pyrazine)-tetrakis(*N,N*-bis(2-hydroxyethyl)dithiocarbamato)dizinc(II) and its di-dioxane solvate. *Z. Krist.* **2017**, *232*, 287–298. [CrossRef]

61. Benson, R.E.; Ellis, C.A.; Lewis, C.E.; Tiekink, E.R.T. 3D-, 2D- and 1D-supramolecular structures of {Zn[$S_2CN(CH_2CH_2OH)R$]$_2$}$_2$ and their {Zn[$S_2CN(CH_2CH_2OH)R$]$_2$}$_2$(4,4'-bipyridine) adducts for R = CH_2CH_2OH, Me or Et: Polymorphism and pseudo-polymorphism. *CrystEngComm* **2007**, *9*, 930–940. [CrossRef]

62. Tan, Y.S.; Tiekink, E.R.T. Crystal structure of (4,4'-bipyridyl-κ*N*)bis[*N*-(2-hydroxyethyl)-*N*-isopropyldithiocarbamato-κ^2S,*S'*]-zinc(II)–4,4'-bipyridyl (2/1) and its isostructural cadmium(II) analogue. *Acta Crystallogr. E Crystallogr. Commun.* **2017**, *73*, 1642–1646. [CrossRef] [PubMed]

63. Broker, G.A.; Jotani, M.M.; Tiekink, E.R.T. Bis[*N*-2-hydroxyethyl,*N*-methyldithiocarbamato-κ^2S,*S*)'-4-{[(pyridin-4-ylmethylidene)hydrazinylidene}methyl]pyridine-κ*N*^{1})zinc(II): Crystal structure and Hirshfeld surface analysis. *Acta Crystallogr. E Crystallogr. Commun.* **2017**, *73*, 1458–1564. [CrossRef] [PubMed]

64. Poplaukhin, P.; Tiekink, E.R.T. Interwoven coordination polymers sustained by tautomeric forms of the bridging ligand. *CrystEngComm* **2010**, *12*, 1302–1306. [CrossRef]

65. Poplaukhin, P.; Arman, H.D.; Tiekink, E.R.T. Supramolecular isomerism in coordination polymers sustained by hydrogen bonding: Bis[Zn($S_2CN(Me)CH_2CH_2OH$)$_2$](*N,N'*-bis(pyridin-3-ylmethyl)thioxalamide). *Z. Krist.* **2012**, *227*, 363–368. [CrossRef]

66. Arman, H.D.; Poplaukhin, P.; Tiekink, E.R.T. Bis[*N*-(2-hydroxyethyl)-*N*-methyldithiocarbamato-κ*S*][2,4,6-tris (pyridin-2-yl)-1,3,5-triazine-κ^3N^1,N^2,N^6]zinc dioxane sesquisolvate. *Acta Crystallogr. E Crystallogr. Commun.* **2012**, *68*, m319–m320. [CrossRef]

67. Ikeda, T.; Hagihara, H. The crystal structure of zinc ethylxanthate. *Acta Crystallogr.* **1966**, *21*, 919–927. [CrossRef]

68. Lai, C.S.; Lim, Y.X.; Yap, T.C.; Tiekink, E.R.T. Molecular paving with zinc thiolates. *CrystEngComm* **2002**, *4*, 596–600. [CrossRef]

69. Addison, A.W.; Rao, T.N.; Reedijk, J.; van Rijn, J.; Verschoor, G.C. Synthesis, structure, and spectroscopic properties of copper(II) compounds containing nitrogen-sulphur donor ligands; the crystal and molecular structure of aqua[1,7-bis(N-methylbenzimidazol-2′-yl)-2,6-dithiaheptane]copper(II) perchlorate. *J. Chem. Soc. Dalton Trans.* **1984**, 1349–1356. [CrossRef]

70. Tiekink, E.R.T. Aggregation patterns in the crystal structures of organometallic Group XV 1,1-dithiolates: The influence of the Lewis acidity of the central atom, metal- and ligand-bound steric bulk, and coordination potential of the 1,1-dithiolate ligands upon supramolecular architecture. *CrystEngComm* **2006**, *8*, 104–118. [CrossRef]

71. Jotani, M.M.; Zukerman-Schpector, J.; Sousa Madureira, L.; Poplaukhin, P.; Arman, H.D.; Miller, T.; Tiekink, E.R.T. Structural, Hirshfeld surface and theoretical analysis of two conformational polymorphs of N,N′-bis(pyridin-3-ylmethyl)oxalamide. *Z. Krist.* **2016**, *231*, 415–425. [CrossRef]

72. Zukerman-Schpector, J.; Sousa Madureira, L.; Poplaukhin, P.; Arman, H.D.; Miller, T.; Tiekink, E.R.T. Conformational preferences for isomeric N,N′-bis(pyridin-n-ylmethyl)ethanedithiodiamides, n = 2, 3 and 4: A combined crystallographic and DFT study. *Z. Krist.* **2015**, *230*, 531–541. [CrossRef]

73. Jamaludin, N.S.; Halim, S.N.A.; Khoo, C.-H.L.; Chen, B.-J.; See, T.-H.L.; Sim, J.-H.; Cheah, Y.-K.; Seng, H.-L.; Tiekink, E.R.T. Bis(phosphane)copper(I) and silver(I) dithiocarbamates: Crystallography and anti-microbial assay. *Z. Krist.* **2016**, *231*, 341–349. [CrossRef]

74. Abrahams, B.F.; Hoskins, B.F.; Winter, G. The structure of cadmium bis(isopropylxanthate)-4,4′-bipyridine. *Aust. J. Chem.* **1990**, *43*, 1759–1765. [CrossRef]

75. Li, T.; Li, Z.-H.; Du, S.-W. *catena*-Poly[[bis(O,O′-dimethyl dithiophosphato-κ²S,S′)cadmium(II)]-μ-4,4′-bipyridine-N:N′]. *Acta Crystallogr. E Crystallogr. Commun.* **2005**, *61*, m95–m97. [CrossRef]

76. Li, T.; Li, Z.-H.; Du, S.-W. *catena*-Poly[[bis(O,O′-diethyl dithiophosphato-κ²S,S′)cadmium(II)]-μ-4,4′-bipyridine-N:N′]. *Acta Crystallogr. E Crystallogr. Commun.* **2004**, *60*, m1912–m1914. [CrossRef]

77. Lai, C.S.; Tiekink, E.R.T. Engineering polymers with variable topology—Bipyridine adducts of cadmium dithiophosphates. *CrystEngComm* **2004**, *6*, 593–605. [CrossRef]

78. Lai, C.S.; Tiekink, E.R.T. Polymeric topologies in cadmium(II) dithiophosphate adducts of the isomeric n-pyridinealdazines, n = 2, 3 and 4. *Z. Krist.* **2006**, *221*, 288–293. [CrossRef]

79. Lai, C.S.; Tiekink, E.R.T. Delineating the principles controlling polymer formation and topology in zinc(II)- and cadmium(II)-dithiophosphate adducts of diimine-type ligands. *J. Mol. Struct.* **2006**, *796*, 114–118. [CrossRef]

80. Devillanova, F.; Aragoni, C.; Arca, M.; Huth, S.L.; Hursthouse, M.B. CAGLIARI 143-$C_{26}H_{28}Cd_1N_2O_4P_2S_4$. *Pers. Commun. Camb. Str. Database* **2008**. [CrossRef]

81. Devillanova, F.; Aragoni, C.; Arca, M.; Hursthouse, M.B.; Huth, S.L. CAGLIARI 161-$C_{28}H_{32}Cd_1N_2O_4P_2S_4$. *Pers. Commun. Camb. Str. Database* **2008**. [CrossRef]

82. Devillanova, F.; Aragoni, C.; Arca, M.; Huth, S.L.; Hursthouse, M.B. CAGLIARI 145-$C_{28}H_{30}Cd_1N_2O_4P_2S_4$. *Pers. Commun. Camb. Str. Database* **2008**. [CrossRef]

83. Devillanova, F.; Aragoni, C.; Arca, M.; Huth, S.L.; Hursthouse, M.B. CAGLIARI 144a-$C_{28}H_{32}Cd_1N_2O_4P_2S_4$. *Pers. Commun. Camb. Str. Database* **2008**. [CrossRef]

84. Devillanova, F.; Aragoni, C.; Arca, M.; Hursthouse, M.B.; Huth, S.L. CAGLIARI 142-$C_{29}H_{34}Cd_1N_2O_4P_2S_4$. *Pers. Commun. Camb. Str. Database* **2008**. [CrossRef]

85. Larionov, S.V.; Shchukin, V.G.; Glinskaya, L.A.; Klevtsova, R.F.; Mazhara, A.P. Mixed-ligand coordination compounds CdL{(i-C₄H₉)₂PS₂}₂ (L = phen, 2,2′-Bipy, and 4,4′-Bipy): Synthesis, structure, and thermal properties. Polymeric structure of [Cd(4,4′-Bipy){(i-C₄H₉)₂PS₂}₂]ₙ. *Russ. J. Coord. Chem.* **2001**, *27*, 463–468. [CrossRef]

86. Fan, J.; Wei, F.-X.; Zhang, W.-G.; Yin, X.; Lai, C.S.; Tiekink, E.R.T. Small ligands-induced synthesis of cadmium complexes with N,N-dibenzyl dithiocarbamate and their crystal structures. *Acta Chim. Sin.* **2007**, *65*, 2014–2018.

87. Chai, J.; Lai, C.S.; Yan, J.; Tiekink, E.R.T. Polymeric [bis(N,N-diethyldithiocarbamato) (trans-1,2-bis(4-pyridyl)ethylene)cadmium(II)]. *Appl. Organomet. Chem.* **2003**, *17*, 249–250. [CrossRef]

88. Avila, V.; Benson, R.E.; Broker, G.A.; Daniels, L.M.; Tiekink, E.R.T. *catena*-Poly[[bis(*N,N*-diethyldithiocarbamato-κ²*S,S′*)cadmium(II)]-μ-trans-1,2-di-4-pyridylethane-κ²*N:N′*]. *Acta Crystallogr. E Crystallogr. Commun.* **2006**, *62*, m1425–m1427. [CrossRef]

89. Poplaukhin, P.; Tiekink, E.R.T. (μ-2-Pyridinealdazine-κ⁴*N,N′:N″,N‴*)bis[bis(*N,N*-di-n-propyldithiocarbamato-κ²*S,S′*)cadmium(II)]. *Acta Crystallogr. E Crystallogr. Commun.* **2008**, *64*, m1176. [CrossRef]

90. Jotani, M.M.; Poplaukhin, P.; Arman, H.D.; Tiekink, E.R.T. [μ₂-trans-1,2-Bis(pyridin-4-yl)ethene-κ²*N:N′*] bis{[1,2-bis(pyridin-4-yl)ethene-κ*N*]bis[*N*-(2-hydroxyethyl)-*N*-isopropyldithiocarbamato-κ²*S,S′*]cadmium} acetonitrile tetrasolvate: Crystal structure and Hirshfeld surface analysis. *Acta Crystallogr. E Crystallogr. Commun.* **2016**, *72*, 1085–1092. [CrossRef] [PubMed]

91. Broker, G.A.; Tiekink, E.R.T. Bis[*N*-(2-hydroxyethyl)-*N*-propyldithiocarbamato-κ²*S,S′*]bis(4-{[(pyridin-4-ylmethylidene)hydrazinylidene]methyl}pyridine-κ*N¹*)cadmium. *Acta Crystallogr. E Crystallogr. Commun.* **2011**, *67*, m320–m321.

92. Arman, H.D.; Poplaukhin, P.; Tiekink, E.R.T. An unprecedented binuclear cadmium dithiocarbamate adduct: Bis[μ₂-*N*-(2-hydroxyethyl)-*N*-isopropylcarbamodithioato-κ³*S*:*S,S′*]bis{[*N*-(2-hydroxyethyl)-*N*-isopropylcarbamodithioato-κ²*S,S′*](3-{(1E)-[(E)-2-(pyridin-3-ylmethylidene)hydrazin-1-ylidene]methyl}-pyridine-κ*N*)cadmium} dihydrate. *Acta Crystallogr. E Crystallogr. Commun.* **2016**, *72*, 1234–1238. [CrossRef] [PubMed]

93. Tan, Y.S.; Sudlow, A.L.; Molloy, K.C.; Morishima, Y.; Fujisawa, K.; Jackson, W.J.; Henderson, W.; Halim, S.N.B.A.; Ng, S.W.; Tiekink, E.R.T. Supramolecular isomerism in a cadmium bis(*N*-hydroxyethyl, *N*-isopropyldithiocarbamate) compound: Physiochemical characterization of ball (*n* = 2) and chain (*n* = ∞) forms of {Cd[S₂CN(iPr)CH₂CH₂OH]₂·solvent}ₙ. *Cryst. Growth Des.* **2013**, *13*, 3046–3056. [CrossRef]

94. Tan, Y.S.; Halim, S.N.A.; Tiekink, E.R.T. Exploring the crystallization landscape of cadmium bis(*N*-hydroxyethyl, *N*-isopropyl-dithiocarbamate), Cd[S₂CN(iPr)CH₂CH₂OH]₂. *Z. Krist.* **2016**, *231*, 113–126. [CrossRef]

crystals

MDPI

Article

Observations on Nanoscale Te Precipitates in CdZnTe Crystals Grown by the Traveling Heater Method Using High Resolution Transmission Electron Microscopy

Boru Zhou [1,2,*]**, Wanqi Jie** [1,2]**, Tao Wang** [1,2]**, Zongde Kou** [1]**, Dou Zhao** [1,2]**, Liying Yin** [1,2]**, Fan Yang** [1,2]**, Shouzhi Xi** [1,2]**, Gangqiang Zha** [1,2] **and Ziang Yin** [1,2]

[1]   State Key Laboratory of Solidification Processing, Northwestern Polytechnical University, Xi'an 710072, China; jwq@nwpu.edu.cn (W.J.); taowang@nwpu.edu.cn (T.W.); kezideanm@126.com (Z.K.); z.dou@mail.nwpu.edu.cn (D.Z.); yinliying324@gmail.com (L.Y.); yang2012100392@163.com (F.Y.); xiaola507@163.com (S.X.); zha_gq@nwpu.edu.cn (G.Z.); yindy90@mail.nwpu.edu.cn (Z.Y.)

[2]   Key Laboratory of Radiation Detection Materials and Devices, Ministry of Industry and Information Technology, School of Materials Science and Engineering, Northwestern Polytechnical University, Xi'an 710072, China

*   Correspondence: zhoubr@mail.nwpu.edu.cn; Tel.: +86-29-8846-0445

Received: 20 December 2017; Accepted: 8 January 2018; Published: 10 January 2018

Abstract: Te precipitates in CdZnTe (CZT) crystals grown by the traveling heater method (THM) are investigated using high-resolution transmission electron microscopy (HRTEM). The results show that in THM-grown CZT crystals, Te precipitates are less than 10 nm in size—much smaller than those in Bridgman-grown CZT. They have hexagonal structure and form a coherent interface with zinc blend structure CZT matrix in the orientation relationship $[\bar{1}12]_M//[0001]_P$ and $(1\bar{1}1)_M//(\bar{1}100)_P$. A ledge growth interface with the preferred orientation along the $[1\bar{1}1]_M$ and $[110]_M$ was found near Te precipitates. The growth and nucleation mechanism of Te precipitates are also discussed.

Keywords: transmission electron microscopy (TEM); traveling heater method; precipitation; interface structure; defects in semiconductors; CdZnTe

1. Introduction

Te precipitates are stubborn defects in CdZnTe (CZT) crystals which degrade the optical transmission and carrier transport properties, and restrict the application of CZT crystal for variant devices [1–3]. Te precipitates are formed by collecting super-saturated point defects such as Te anti-sites, Te interstitials, or cation vacancies, due to the retrograde solubility of Te atoms in the phase diagram. They are normally tens of nanometers in size, which is much smaller than the micrometer-scale Te inclusions formed by trapping Te-rich droplets at the growth interface during the growth process. Te inclusions have already been widely studied in the aspects of morphology evolution, formation mechanism, the effects on CZT-base device performance, etc. [1,4–7]. However, nano-scale Te precipitate seems to be more complicated and is not well understood.

The formation of Te precipitates is also thought to be related to the crystal growth method. For the popular Bridgman growth, the growth temperature is high (at the melting point of the crystal), which will produce a high density of point defects and promote the formation of Te precipitates. Rai et al. [8] characterized the density and size of Te precipitates in CZT crystals grown by Bridgman method. Te precipitates were identified to possess an average size of 20 nm and a density of 1.3×10^{16} cm^{-3} in the as-grown CZT crystal. Wang et al. [9] further studied the crystal structure of Te precipitates in

CZT crystals, and three structure types (i.e., hexagonal, monoclinic, and high-pressure rhombohedral structures) were characterized.

The traveling heater method (THM) is believed to be one of most promising methods for growing high-quality CZT crystals [10] due to the lower growth temperature. Referring to the Cd-Te phase diagram and projections of $Cd_{0.9}Zn_{0.1}Te$ solidus along the composition axis [11,12], THM growth in Te solution will decrease the growth temperature, which is favorable for lowering the solid solubility of Te atoms in as-grown CZT crystals and decreasing the possibility of Te precipitate formation.

In this work, we focus on the experimental observations of nano-scale Te precipitates in CZT crystals grown by THM using high-resolution transmission electron microscopy (HRTEM) to reveal the behaviors and underlying mechanisms of Te precipitates associated with crystal growth conditions in THM process.

2. Materials and Methods

$Cd_{0.9}Zn_{0.1}Te$ crystals were grown by THM (self-designed, Xi'an, China) with the accelerated crucible rotation technique (ACRT, self-designed, Xi'an, China), as described in our previous work [13]. After the growth, the ingots were cooled to room temperature at two different cooling rates (5 °C/h and 60 °C/h). For HRTEM analyses, several single crystal specimens with the dimensions $2.5 \times 2.5 \times 2$ mm^3 were cut from the ingots by a diamond wire cutting machine. These specimens were mechanically ground to a thickness of 20–30 μm, and then thinned by Ar$^+$ ion in a plasma etching equipment (PIPC-691, Gatan, Shanghai, China) until an electron transparent thin area was acquired. A Tecnai F30 G2 transmission electron microscope (FEI, Hillsboro, OR, United States) with a spatial resolution of 0.2 nm and electron energy of 300 keV was employed to analyze the nanoscale Te precipitates and their orientation relationship with CZT matrix. For the orientation relationship analysis, electron beam diffraction method was used to confirm the results.

3. Results and Discussion

Figure 1a shows the bright-field TEM micrograph of a CZT sample. Even though Te precipitates are not visible, the selected area electron diffraction (SAED) pattern in Figure 1b shows two sets of diffraction spots. The strong diffraction spots (marked by the red rhomboid) correspond to CZT matrix phase with the zinc-blende structure (PDF number: 50–1440, space group F-43m, lattice constant 0.6456 nm) along the $[\bar{1}12]_M$ zone axis. The weak spots (marked by the yellow rhomboid) are identified as Te precipitate phases with the hexagonal structure along the $[0001]_P$ zone axis (PDF number: 36–1452, space group: P3121, lattice constants: a = 0.4458 nm and c = 0.5927 nm). Here, the subscripts M and P represent CZT matrix and Te precipitates, respectively. The orientation relationship between Te precipitates and CZT matrix is derived to be $[\bar{1}12]_M//[0001]_P$ and $(1\bar{1}1)_M//(\bar{1}100)_P$. Under hydrostatic pressure, element Te may undergo pressure-induced phase transitions from trigonal to monoclinic at about 40 kbar and from monoclinic to rhombohedral phase at about 70 kbar, respectively [14–16]. Final structures of Te precipitates are therefore decided by the internal droplet pressures. We saw only hexagonal structure in THM-grown CZT crystals, which implies that Te precipitates can be formed at a pressure less than 40 kbar or due to the confinement of CZT matrix.

Figure 1c shows the dark-field TEM micrograph corresponding to the weak diffraction spot $(0\bar{1}10)_P$. It shows that Te precipitates distribute homogeneously in CZT crystals. Different from 10–30 nm Te precipitates in Bridgman-grown CZT crystals [1,8,9], in our THM-grown CZT crystals, Te sizes were smaller than 10 nm and the density was about 9.2×10^{15} cm^{-3}. The results demonstrate that THM with lower growth temperature can markedly reduce the size of Te precipitates in CZT crystals, which is favorable for improving the photoelectric properties and the structural perfection of CZT crystals.

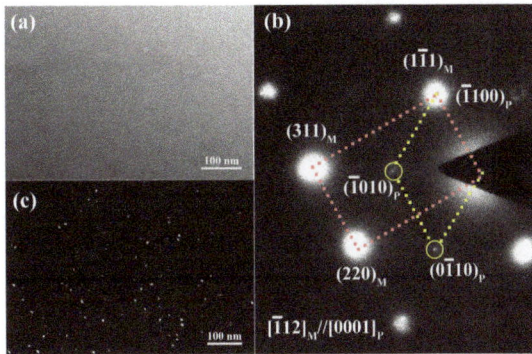

Figure 1. Micrographs and selected area electron diffraction (SAED) pattern of CdZnTe (CZT) crystals grown by the traveling heater method (THM) with slow cooling rate along the $[\bar{1}12]_M$ zone axis. (a) Bright-field transmission electron microscopy (TEM) micrograph along the $[\bar{1}12]_M$ zone axis. (b) Dark-field TEM micrograph along the $[\bar{1}12]_M$ zone axis. (c) SAED pattern along the $[\bar{1}12]_M$ zone axis.

The HRTEM images of CZT sample along the $[\bar{1}12]_M$ zone axis are shown in Figure 2. The interfaces between Te precipitates and CZT matrix were approximately outlined according to the inverse fast Fourier transform (IFFT) image. Hexagonal Te precipitates smaller than 10 nm in size were observed, shown in Figure 2a. A fast Fourier transform (FFT) pattern of region I in Figure 2a is shown in the inset of Figure 2a,b, respectively. The IFFT image was reconstructed by using the $(\bar{1}100)_P$ spatial frequencies as marked by red circles in the FFT image, shown in Figure 2b. No lattice mismatch was found at the interface between Te precipitates and CZT matrix. This indicates a perfect coherent relationship, which may be the reason for hardly observing Te precipitates in the bright-field TEM micrograph (Figure 1a).

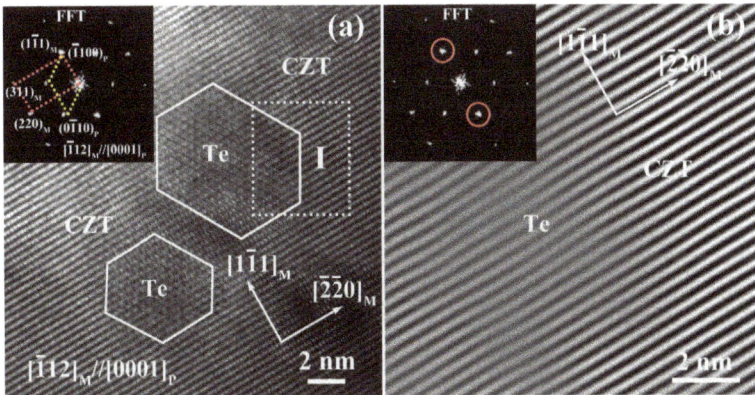

Figure 2. High-resolution TEM (HRTEM) images of Te precipitates along $[\bar{1}12]_M$ zone axis. (a) The HRTEM image, (b) The inversed fast Fourier transform (IFFT) image of the region I in (a). The two insets are FFT patterns of region I in (a).

Figure 3a shows an HRTEM image of the specimen near the ingot tail with fast cooling rate (60 °C/h) after growth, where Te precipitates in the irregular and faceted morphology were seen. A ledge growth interface was found on the edge of Te precipitates, which are parallel to the $[1\bar{1}1]_M$

direction and perpendicular to the $[220]_M$ direction of the CZT matrix. Therefore, the optimal growth direction of Te precipitates is along $[1\bar{1}1]_M$ and $[110]_M$ in the CZT matrix, attributed to the low interfacial energies on $\{111\}_M$ and $\{110\}_M$ planes between CZT matrix and Te precipitate phases [17,18]. The faceted morphology of Te precipitates can be caused by a ledge growth mechanism. It is possible that the as-grown interface between Te precipitates and CZT matrix can be retained by the fast cooling rate (60 °C/h) after the growth. The irregular and faceted interface morphology (Figure 3a) indicates that the as-grown interface was obtained. Different from the coherent interface (Figure 2b) under the slowly cooling condition (5 °C/h), the dislocations labeled D and the stacking fault labeled S are seen near the as-grown interface between Te precipitates and CZT matrix along $[110]_M$ growth direction (Figure 3b). This indicates that it is possible that the dislocations can be main channels for the diffusion and transport of Te atoms in CZT matrix, which play a crucial role in the growth process of Te precipitates in CZT matrix during cooling down.

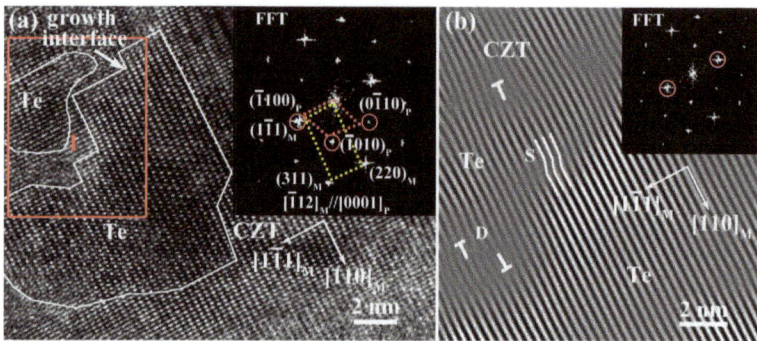

Figure 3. HRTEM image of the specimen near the tail of the ingot with fast cooling rate (60 °C/h) after growth. (**a**) The HRTEM image. (**b**) The inverse FFT image was reconstructed by using the $(\bar{1}100)_P$ spatial frequencies shown in the inset of (**b**). The insets in (**a**,**b**) correspond to FFT patterns of region I in (**a**).

The formation process of Te precipitates is further illustrated in Figure 4. Te precipitates originated from the native point defect condensation [19]. Principal defects are Te interstitials, Cd vacancies, and Te anti-sites in CZT crystals when grown by THM from Te solution. Te_{Cd} anti-sites with the high migration energy barrier (1.68 eV) are almost immovable, while Te_i interstitials with a very low barrier (0.16 eV) are extremely movable along the $[110]_M$ direction [20], which is consistent with the optimal growth direction of Te precipitates shown in Figure 3a. The migration energy barrier of V_{Cd} vacancy is 1.09 eV, between those of Te_{Cd} antisites and Te_i interstitials [20]. Therefore, Te interstitials can be the dominant diffusing species in the nucleation and growth processes of Te precipitates. After the growth, the oversaturated Te interstitial atoms aggregated to form Te precipitation nuclei due to the retrograde solidus of Te atoms in CZT matrix during the cooling of CZT crystals, as shown in Figure 4a. These nuclei caused the lattice distortions, resulting in the formation of dislocations around the nuclei. The dislocations can provide a fast channel for the diffusion and transport of Te interstitial atoms, which promote the growth of the nuclei until the whole excess Te-related point defects (priority to Te interstitials) have precipitated, as shown in Figure 4b. The real shapes of Te precipitates always mean to minimize the elastic strain energy and the total interfacial energy in order to lower the total Gibbs energy. Coherent Te precipitates have been observed in CZT matrix (Figure 2), indicating that the lattice strain energy may be very low. For nano-size Te precipitates, the interfacial energy may contribute more to the shapes of the precipitates for lowering the total system energy [21]. According to theoretical results [17], {111} faces have the lowest interfacial energy in zinc-blende crystal structure. Therefore, the equilibrium morphology of Te precipitates can be bounded by $\{111\}_M$ faces and forms

Crystals **2018**, *8*, 26

a regular polygon. Therefore, the corresponding Te precipitates have also been observed, as shown in Figure 4c.

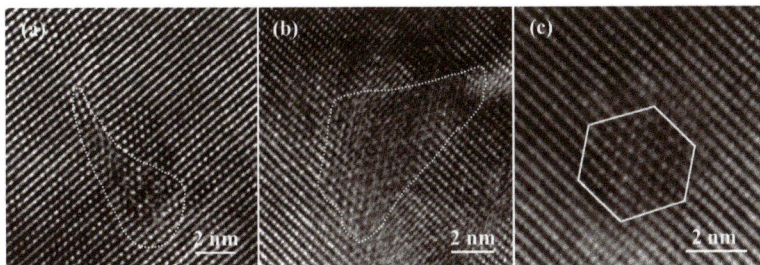

Figure 4. The formation process of Te precipitates: (**a**) nucleation; (**b**) growth; (**c**) the final morphology along the $[\bar{1}12]_M$ zone axis.

4. Conclusions

Nanoscale Te precipitates in CZT crystals grown by THM were observed using HRTEM. The majority of Te precipitates were found to be smaller than 10 nm in size with a density of 9.2×10^{15} cm^{-3}—much smaller than those in Bridgman-grown CZT crystals. This demonstrates that THM with lower growth temperature is an effective way to reduce tendency to Te precipitation in CZT crystals. Moreover, Te precipitates with a hexagonal structure have a coherent interface with the CZT matrix. The orientation relationship between CZT with zinc blend structure and Te precipitates with hexagonal structure is $[\bar{1}12]_M // [0001]_P$ and $(1\bar{1}1)_M // (\bar{1}100)_P$. In addition, a ledge growth interface was found on the edge of Te precipitates, and the preferred growth directions were along $[1\bar{1}1]_M$ and $[110]_M$ directions in the CZT matrix, attributed to their low interfacial energies.

Acknowledgments: This work was supported by the National Key Research and Development Program of China (2016YFB0402405, 2016YFF0101301), the National Natural Science Foundation of China (51672216, NNSFC-51502244), the Fundamental Research Funds for the Central Universities (3102016ZY011, 3102015BJ(II)ZS014), the Research Fund of the State Key Laboratory of Solidification Processing (NWPU), China (94-QZ-2014).

Author Contributions: Boru Zhou performed the experiments, analyzed the results and wrote the manuscript. Wanqi Jie and Tao Wang led the project, conceived the ideas, and revised the manuscript. Zongde Kou analyzed the results from the selected area electron diffraction (SAED) pattern. Dou Zhao prepared TEM specimens. Liying Yin, Fan Yang, Shouzhi Xi, Gangqiang Zha and Ziang Yin helped the manuscript and the interpretation of the results.

Conflicts of Interest: The authors declare no conflict of interest.

References

1. Rudolph, P.; Engel, A.; Schentke, I.; Grochocki, A. Distribution and genesis of inclusions in CdTe and (Cd, Zn) Te single crystals grown by the Bridgman method and by the travelling heater method. *Cryst. Growth* **1995**, *147*, 297–304. [CrossRef]

2. Carini, G.A.; Bolotnikov, A.E.; Camarda, G.S.; Wright, G.W.; James, R.B.; Li, L. Effect of Te precipitates on the performance of CdZnTe detectors. *Appl. Phys. Lett.* **2006**, *88*, 143515. [CrossRef]

3. Zhu, J.; Zhang, X.; Li, B.; Chu, J. The effects of Te precipitation on IR transmittance and crystalline quality of as-grown CdZnTe crystals. *Infrared Phys. Technol.* **1999**, *40*, 411–415. [CrossRef]

4. He, Y.; Jie, W.; Xu, Y.; Wang, T.; Zhang, G.; Yu, P.; Zheng, X.; Zhou, Y.; Liu, H. Matrix-controlled morphology evolution of Te inclusions in CdZnTe single crystal. *Scr. Mater.* **2012**, *67*, 5–8. [CrossRef]

5. Bolotnikov, A.E.; Abdul-Jabbar, N.M.; Babalola, O.S.; Camarda, G.S.; Cui, Y.; Hossain, A.M.; Jackson, E.M.; Jackson, H.C.; James, J.A.; Kohman, K.T.; et al. Effects of Te inclusions on the performance of CdZnTe radiation detectors. *IEEE Trans. Nucl. Sci.* **2008**, *55*, 2757–2764. [CrossRef]

6. Marchini, L.; Zappettini, A.; Zha, M.; Zambelli, N.; Bolotnikov, A.E.; Camarda, G.S.; James, R.B.; Marchini, L.; Zappettini, A.; Zha, M.; et al. Crystal defects in CdZnTe crystals grown by the modified low-pressure Bridgman method. *IEEE Trans. Nucl. Sci.* **2012**, *59*, 264–267. [CrossRef]

7. Roy, U.N.; Weiler, S.; Stein, J.; Hossain, A.; Camarda, G.S.; Bolotnikov, A.E.; James, R.B. Size and distribution of Te inclusions in THM as-grown CZT wafers: The effect of the rate of crystal cooling. *Cryst. Growth* **2011**, *332*, 34–38. [CrossRef]

8. Rai, R.S.; Mahajan, S.; McDevitt, S.; Johnson, C.J. Characterization of CdTe (Cd, Zn) Te, and Cd (Te, Se) single crystals by transmission electron microscopy. *Vac. Sci. Technol. B* **1991**, *235*, 1892–1896. [CrossRef]

9. Wang, T.; Jie, W.; Zeng, D. Observation of nano-scale Te precipitates in cadmium zinc telluride with HRTEM. *Mater. Sci. Eng. A* **2008**, *472*, 227–230. [CrossRef]

10. Chen, H.; Awadalla, S.A.; Iniewski, K.; Lu, P.H.; Harris, F.; Mackenzie, J.; Hasanen, T.; Chen, W.; Redden, R.; Bindley, G.J. Characterization of large cadmium zinc telluride crystals grown by traveling heater method. *J. Appl. Phys.* **2008**, *103*, 1–5. [CrossRef]

11. Haloui, A.; Feutelais, Y.; Legendre, B. Experimental study of the ternary system Cd Te Zn. *J. Alloys Compd.* **1997**, *260*, 179–192. [CrossRef]

12. Greenberga, J.H.; Guskovb, V.N. Vapor pressure scanning of non-stoichiometry in $Cd_{0.9}Zn_{0.1}Te_{1\pm\delta}$ and $Cd_{0.85}Zn_{0.15}Te_{1\pm\delta}$. *J. Cryst. Growth* **2006**, *289*, 552–558.

13. Zhou, B.; Jie, W.; Wang, T.; Xu, Y.; Yang, F.; Yin, L.; Zhang, B.; Nan, R. Growth and Characterization of Detector-Grade $Cd_{0.9}Zn_{0.1}Te$ Crystals by the Traveling Heater Method with the Accelerated Crucible Rotation Technique. *J. Electron. Mater.* **2017**, 1–6. [CrossRef]

14. Jamieson, J.C.; McWhan, D.B. Crystal structure of tellurium at high pressures. *J. Chem. Phys.* **1965**, *43*, 1149–1152. [CrossRef]

15. Aoki, K.; Shimomura, O.; Minomura, S. Crystal Structure of the High-Pressure Phase of Tellurium. *J. Phys. Soc. Jpn.* **1980**, *48*, 551–556. [CrossRef]

16. Yadava, R.D.S.; Bagai, R.B.; Borle, W.N. Theory of Te precipitation and related effects in CdTe Crystals. *J. Electron. Meter.* **1992**, *21*, 1001–1016. [CrossRef]

17. Hu, S.; Henager, C.H., Jr. Phase-field simulations of Te-precipitate morphology and evolution kinetics in Te-rich CdTe crystals. *J. Cryst. Growth* **2009**, *311*, 3184–3194. [CrossRef]

18. Caginalp, G.; Fife, P. Higher-order phase field models and detailed anisotropy. *Phys. Rev. B* **1986**, *34*, 4940–4943. [CrossRef]

19. Rudolph, P. Fundamentals and engineering of defects. *Prog. Cryst. Growth Charact.* **2016**, *62*, 89–110. [CrossRef]

20. Lordi, V. Point defects in Cd(Zn)Te and TlBr: Theory. *J. Cryst. Growth* **2013**, *379*, 84–92. [CrossRef]

21. Solomon, E.L.S.; Araullo-Peters, V.; Allison, J.E.; Marquis, E.A. Early precipitate morphologies in Mg-Nd-(Zr). *Alloys. Scr. Mater.* **2017**, *128*, 14–17. [CrossRef]

crystals

MDPI

Article

Crystal Chemistry of Zinc Quinaldinate Complexes with Pyridine-Based Ligands

Barbara Modec

Department of Chemistry and Chemical Technology, University of Ljubljana; Večna pot 113, 1000 Ljubljana, Slovenia; barbara.modec@fkkt.uni-lj.si; Tel.: +386-1-4798-526

Received: 1 December 2017; Accepted: 16 January 2018; Published: 19 January 2018

Abstract: Substitution of methanol in $[Zn(quin)_2(CH_3OH)_2]$ ($quin^-$ denotes an anionic form of quinoline-2-carboxylic acid, also known as quinaldinic acid) with pyridine (Py) or its substituted derivatives, 3,5-lutidine (3,5-Lut), nicotinamide (Nia), 3-hydroxypyridine (3-Py-OH), 3-hydroxymethylpyridine (3-Hmpy), 4-hydroxypyridine (4-Py-OH) and 4-hydroxymethylpyridine (4-Hmpy), afforded a series of novel heteroleptic complexes with compositions $[Zn(quin)_2(Py)_2]$ (**1**), $[Zn(quin)_2(3,5-Lut)_2]$ (**2**), $[Zn(quin)_2(Nia)_2] \cdot 2CH_3CN$ (**3**), $[Zn(quin)_2(3-Py-OH)_2]$ (**4**), $[Zn(quin)_2(3-Hmpy)_2]$ (**5**), $[Zn(quin)_2(4-Pyridone)]$ (**6**) (4-Pyridone = a keto tautomer of 4-hydroxypyridine), and $[Zn(quin)_2(4-Hmpy)_2]$ (**7**). In all reactions, the $\{Zn(quin)_2\}$ structural fragment with quinaldinate ions bound in a bidentate chelating manner retained its structural integrity. With the exception of $[Zn(quin)_2(4-Pyridone)]$ (**6**), all complexes feature a six-numbered coordination environment of metal ion that may be described as a distorted octahedron. The arrangement of ligands is *trans*. The coordination sphere of zinc(II) in the 4-pyridone complex consists of only three ligands, two quinaldinates, and one secondary ligand. The metal ion thereby attains a five-numbered coordination environment that is best described as a distorted square-pyramid (τ parameter equals 0.39). The influence of substituents on the pyridine-based ligand over intermolecular interactions in the solid state is investigated. Since pyridine and 3,5-lutidine are not able to form hydrogen-bonding interactions, the solid state structures of their complexes, $[Zn(quin)_2(Py)_2]$ (**1**) and $[Zn(quin)_2(3,5-Lut)_2]$ (**2**), are governed by $\pi \cdots \pi$ stacking, C–H$\cdots \pi$, and C–H$\cdots$O intermolecular interactions. With other pyridine ligands possessing amide or hydroxyl functional groups, the connectivity patterns in the crystal structures of their complexes are governed by hydrogen bonding interactions. Thermal decomposition studies of novel complexes have shown the formation of zinc oxide as the end product.

Keywords: zinc(II) complexes; quinaldinic acid; pyridine; hydroxyl group; crystal structure; hydrogen bond

1. Introduction

The main interest in coordination chemistry of zinc finds its origin in its biological importance. In 1940, carbonic anhydrase II was discovered by Keilin and Mann as the first zinc-containing metalloenzyme [1]. The enzyme plays a key role in the transformation of carbon dioxide into bicarbonate in blood or in the reverse reaction in lungs. Since then, over 1000 zinc metalloenzymes, covering all classes of enzymes, have been discovered [2–6]. The prevalent use of zinc in biological systems is due to its unique properties [3,5,7–14]. Zinc(II) ion is characterized by a filled *d* subshell. Because of the latter, it lacks any redox activity which makes it an ideal metal cofactor for reactions that require a redox-stable center. The function of the ion is structural or that of a Lewis acid-type catalyst [15]. Since it does not act consistently either as a soft or as a hard Lewis acid, it presents a borderline case with no strong preference for coordination of either oxygen-, nitrogen- or sulfur-donor ligands [16]. Another important property, which is again a direct consequence of the filled *d* subshell, is its ligand-field stabilization energy which is zero in all coordination environments [17]. Although no geometry is inherently more stable than another, the structurally characterized zinc enzymes display in most cases a distorted tetrahedral coordination environment of the metal ion [18,19]. Studies of synthetic analogues of enzymes, i.e., model compounds, could help in understanding how the immediate environment of the metal ion in an enzyme modulates its chemistry [20–22].

Quinaldinic acid (IUPAC name: quinoline-2-carboxylic acid), shown in Scheme 1, is a structural analogue of pyridine-2-carboxylic acid (known also as picolinic acid) which has found a wide use in the coordination chemistry of many transition metals. Upon giving off their protons, both acids form anions which possess two functional groups, a quinoline/pyridine nitrogen and a carboxylate, that can bind to transition metal cations. Although the recent literature abounds with structural data on picolinate complexes [23–27], the quinaldinate complexes remain relatively rare [28–32]. The quinaldinate ion can coordinate to metal ions in several ways (Scheme 2). A recent survey of the structurally characterized zinc(II) complexes with quinaldinate produced altogether six hits [33]. In all, the quinaldinate ion adopts the *N,O*-chelating binding mode through nitrogen and carboxylate oxygen, depicted as **a** in Scheme 2 and labelled as 1.101 in Harris notation [34]. One of the complexes, [Zn(quin)$_2$(CH$_3$OH)$_2$] [35], was chosen for its labile methanol ligands to serve as an entry into the coordination chemistry of heteroleptic complexes that would contain apart from quinaldinate also a pyridine-based ligand. The pyridine ligands, used in this study (shown in Scheme 3), differ in the nature of their substituents and their relative positions on the ring. The ability of chosen ligands to participate in hydrogen-bonding is markedly different. The pyridine ligand is thus expected to have a crucial impact over strength and directionality of intermolecular interactions among complex molecules in solid state structures. Herein, we present the syntheses and the crystal structures of a series of zinc(II) complexes with quinaldinate and pyridine-based ligands. Reactions of methanol complex with pyridine-based ligands afforded [Zn(quin)$_2$(Py)$_2$] (**1**), [Zn(quin)$_2$(3,5-Lut)$_2$] (**2**), [Zn(quin)$_2$(Nia)$_2$]·2CH$_3$CN (**3**), [Zn(quin)$_2$(3-Py-OH)$_2$] (**4**), [Zn(quin)$_2$(3-Hmpy)$_2$] (**5**), [Zn(quin)$_2$(4-Pyridone)] (**6**), and [Zn(quin)$_2$(4-Hmpy)$_2$] (**7**).

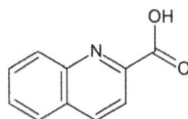

Scheme 1. Structural formula of quinaldinic acid.

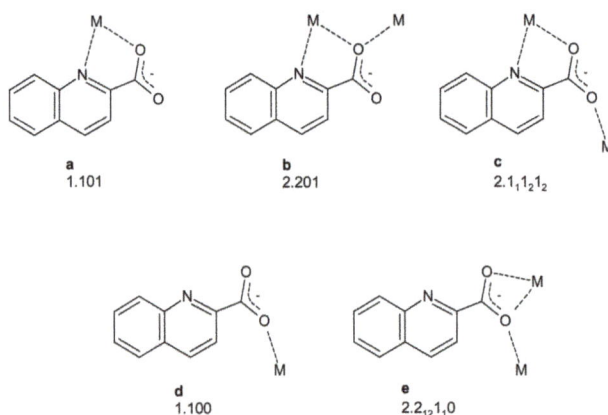

Scheme 2. Various coordination modes of quinaldinate, labelled following the Harris notation [33].

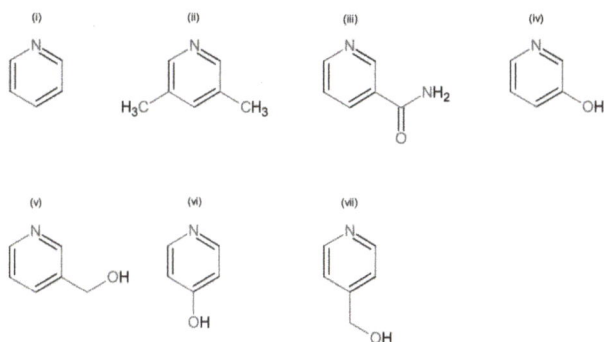

Scheme 3. Pyridine ligands, used in this work: (**i**) pyridine (Py), (**ii**) 3,5-lutidine (3,5-Lut), (**iii**) nicotinamide (Nia), (**iv**) 3-hydroxypyridine (3-Py-OH), (**v**) 3-hydroxymethylpyridine (3-Hmpy), (**vi**) 4-hydroxypyridine (4-Py-OH), and (**vii**) 4-hydroxymethylpyridine (4-Hmpy).

2. Materials and Methods

2.1. General

All manipulations and procedures were conducted in air. With the exception of acetonitrile, chemicals were purchased from Sigma Aldrich (St. Louis, MO, USA) and used as received. Acetonitrile was dried over molecular sieves, following the published procedure [36]. The IR spectra were measured on solid samples using a Perkin Elmer Spectrum 100 series FT-IR spectrometer (PerkinElmer, Shelton, CT, USA), equipped with ATR. Elemental analyses were performed by the Chemistry Department service at the University of Ljubljana. ^{1}H NMR spectra of DMSO-d_6 solutions were recorded on Bruker Avance DPX 300 MHz and/or Bruker Avance III 500 MHz NMR instruments (Bruker BioSpin GmbH, Rheinstetten, Germany). The chemical shifts (δ) were referenced to tetramethylsilane. The choice of dimethyl sulfoxide (DMSO) as a solvent was governed by a meagre solubility of complexes in other solvents. Upon dissolving, a substitution of pyridine-based ligands with DMSO molecules in the zinc(II) coordination sphere took place in all cases. From the solutions in NMR tubes, a complex of zinc(II) with DMSO crystallized as large, block-like crystals of a light yellow color. Their cell dimensions were the same as those of the known compound, [Zn(quin)$_2$(DMSO)$_2$]·2DMSO [37]. The NMR spectra of compounds are thus the spectra of the DMSO complex and *free* pyridine-based ligands. Two representative spectra are given in Supplementary Materials. Thermal analyses were performed on a Mettler Toledo TG/DSC 1 instrument (Mettler Toledo, Schwerzenbach, Switzerland) in argon or air atmosphere at a 50 mL/min gas flow. Masses of the samples were in the 4–32 mg range. Samples in platinum crucibles were heated from room temperature to 800 °C with a heating rate of 5 K/min. In each case, the baseline was subtracted. The solid residue, a mixture of grey amorphous material and zinc oxide (zincite modification), was identified by X-ray powder diffraction (XRPD). XRPD data were collected with a PANalytical X'Pert PRO MD diffractometer (PANALYTICAL, Almelo, The Netherlands) by using Cu-Kα radiation (λ = 1.5406 Å).

2.2. Preparation Procedures

Preparation of [Zn(quin)$_2$(CH$_3$OH)$_2$]. A teflon container was loaded with zinc(II) acetate dihydrate (110 mg, 0.50 mmol) and quinaldinic acid (173 mg, 1.00 mmol). Methanol (15 mL) was added. The container was closed and inserted into a steel autoclave. The autoclave was heated for 24 h at 105 °C. The reaction vessel was then allowed to cool slowly to room temperature. Large colorless crystals of [Zn(quin)$_2$(CH$_3$OH)$_2$] were collected by filtration. The mass of the dried product was 100 mg. Yield: 0.21 mmol, 42%. Note. The crystals were found to lose their lustre after a prolonged period.

Preparation of [Zn(quin)₂(Py)₂] (**1**). A teflon container was loaded with [Zn(quin)₂(CH₃OH)₂] (50 mg, 0.11 mmol), pyridine (2 mL), and dried acetonitrile (7.5 mL). The container was closed and inserted into a steel autoclave. The autoclave was heated for 24 h at 105 °C. The reaction vessel was then allowed to cool slowly to room temperature. Large colorless crystals of **1** were collected by filtration. The mass of the dried product was 41 mg. Yield: 0.072 mmol, 66%. Found C, 63.16; H, 3.86; N, 9.84%. $C_{30}H_{22}N_4O_4Zn$ requires C, 63.45; H, 3.90; N, 9.87%. IR (ATR, cm^{-1}): 3069w, 1647vvs [ν_{asym}(COO)], 1594s, 1566s, 1508m, 1484m, 1467m, 1440vs, 1362vvs, 1351vvs [ν_{sym}(COO)], 1292w, 1270m, 1209m, 1176m, 1159m, 1149m, 1111w, 1071m, 1036m, 1005m, 959m, 891s, 856s, 802vvs, 771vvs, 759vvs, 744s, 703vvs, 636s, 620s. ^{1}H NMR (DMSO) δ/ppm: 7.39 (m, 2H), 7.76–7.85 (m, 1H), 8.58 (m, 2H) [pyridine signals]; 7.76–7.85 (m, 1H), 8.11 (m, 1H), 8.20 (d, J = 7.9 Hz, 1H), 8.41 (d, J = 8.5 Hz, 1H), 8.80 (m, 2H) [quinaldinate signals].

Preparation of [Zn(quin)₂(3,5-Lut)₂] (**2**). A teflon container was loaded with [Zn(quin)₂(CH₃OH)₂] (50 mg, 0.11 mmol), 3,5-lutidine (1 mL), and dried acetonitrile (10 mL). The container was closed and inserted into a steel autoclave. The autoclave was heated for 24 h at 105 °C. The reaction vessel was then allowed to cool slowly to room temperature. Large, light yellow crystals of **2** were collected by filtration. The mass of the dried product was 55 mg. Yield: 0.088 mmol, 80%. Found C, 65.30; H, 4.52; N, 9.02%. $C_{34}H_{30}N_4O_4Zn$ requires C, 65.44; H, 4.85; N, 8.98%. IR (ATR, cm^{-1}): 3113w, 3066w, 2912w, 1654vvs [ν_{asym}(COO)], 1593s, 1564m, 1551m, 1507m, 1458s, 1428m, 1405w, 1355vvs [ν_{sym}(COO)], 1343vs, 1320s, 1294w, 1269m, 1243w, 1215w, 1205w, 1172s, 1147vs, 1115m, 1034w, 1023w, 1000w, 986w, 957s, 940w, 893s, 870s, 861s, 811s, 800vs, 772vvs, 743vs, 706vvs, 635vs. ^{1}H NMR (DMSO) δ/ppm: 2.25 (s, 6H), 7.42 (s, 1 H), 8.19 (s, 2H) [3,5-lutidine signals]; 7.82 (m, 1H), 8.01 (m, 1H), 8.19 (d, J = 7.9 Hz, 1H), 8.41 (d, J = 8.4 Hz, 1H), 8.78 (m, 2H) [quinaldinate signals].

Preparation of [Zn(quin)₂(Nia)₂]·2CH₃CN (**3**). A teflon container was loaded with [Zn(quin)₂(CH₃OH)₂] (50 mg, 0.11 mmol), nicotinamide (120 mg, 0.98 mmol) and dried acetonitrile (10 mL). The container was closed and inserted into a steel autoclave. The autoclave was heated for 24 h at 105 °C. The reaction vessel was then allowed to cool slowly to room temperature. A colorless microcrystalline solid was collected by filtration. Mass of the dried product was 47 mg. Yield: 0.072 mmol, 65%. Note. The crystals lost interstitial solvent upon removal from mother liquor. Found C, 58.28; H, 3.61; N, 12.47%. $C_{32}H_{24}N_6O_6Zn$ (complex without acetonitrile solvent) requires C, 58.77; H, 3.70; N, 12.85%. IR (ATR, cm^{-1}): 3292m, 3147m [ν(N–H), amide], 2256w [ν(CN), acetonitrile], 1692s [ν(C=O), amide], 1676m, 1651w, 1624vvs [ν_{asym}(COO)], 1600vvs, 1565m, 1506w, 1479w, 1459s, 1432m, 1383m, 1363vvs [ν_{sym}(COO)], 1342s, 1330m, 1263w, 1250w, 1217w, 1203m, 1168s, 1159m, 1147s, 1114w, 1059s, 1041w, 1020w, 992w, 957m, 945m, 922m, 897s, 880m, 856s, 807vs, 775vvs, 743m, 695vvs, 670m, 653vs, 632s. ^{1}H NMR (DMSO) δ/ppm: 7.50 (m, 1H), 7.58 (br, 1H), 8.14 (br, 1H), 8.22 (m, 1H), 8.70 (dd, 1H), 9.03 (dd, 1H) [nicotinamide signals]; 7.82(m, 1H), 8.01 (m, 1H), 8.22 (d, J = 7.9 Hz, 1H), 8.41 (d, J = 8.4 Hz, 1H), 8.80 (m, 2H) [quinaldinate signals]; 2.08 (s, 3H) [methyl, acetonitrile].

Preparation of [Zn(quin)₂(3-Py-OH)₂] (**4**). A teflon container was loaded with [Zn(quin)₂(CH₃OH)₂] (50 mg, 0.11 mmol), 3-hydroxypyridine (150 mg, 1.58 mmol), and dried acetonitrile (10 mL). The container was closed and inserted into a steel autoclave. The autoclave was heated for 24 h at 105 °C. The reaction vessel was then allowed to cool slowly to room temperature. The solid consisted mostly of the unreacted ligand (large orange-colored crystals), [Zn(quin)₂(H₂O)] (colorless, crystalline solid) and [Zn(quin)₂(3-Py-OH)₂] (**4**) (a few block-like colorless crystals). The identity of the orange crystalline material was determined by its infrared spectrum and ^{1}H NMR which were identical to the spectra of pure ligand. The aqua complex was identified by its infrared spectrum. Repeated attempts to improve the yield were not met with success.

Preparation of [Zn(quin)₂(3-Hmpy)₂] (**5**). A teflon container was loaded with [Zn(quin)₂(CH₃OH)₂] (50 mg, 0.11 mmol), 3-hydroxymethylpyridine (1.5 mL), and dried acetonitrile (10 mL). The container was closed and inserted into a steel autoclave. The autoclave was heated for 24 h at 105 °C. The reaction vessel was then allowed to cool slowly to room temperature. Large block-like crystals of **5** of a gold yellow color were collected by filtration. Mass of the dried product was 43 mg. Yield: 0.068 mmol, 62%. Found C, 61.27; H, 4.10; N, 9.16%. $C_{32}H_{26}N_4O_6Zn$ requires C, 61.20; H, 4.17; N, 8.92%. IR (ATR, cm^{-1}): 3367s [ν(O–H), 3-Hmpy], 3071w, 2824w, 1638vvs [ν_{asym}(COO)], 1597vs, 1581m, 1566m, 1554m, 1506m, 1479w, 1460s, 1428vs, 1368vvs [ν_{sym}(COO)], 1348m, 1336m, 1305s, 1273m 1234w, 1204s, 1191m, 1177m, 1152s, 1128w, 1107w, 1062vvs, 1048vs [ν(C–O), 3-Hmpy], 1031m, 1010w, 968m, 926w, 897s, 858s, 796vvs, 775vvs, 708vvs, 646m, 633vs. ^{1}H NMR (DMSO) δ/ppm: 4.53 (d, J = 5.6 Hz, 2H), 5.35 (t, J = 5.7 Hz, 1H), 7.37 (m, 1H), 7.73 (d, J = 7.8 Hz, 1H), 8.46 (d, J = 3.7 Hz, 1H), 8.53 (d, J = 1 Hz, 1H) [3-hydroxymethylpyridine signals]; 7.81 (m, 1H), 7.93 (m, 1H), 8.20 (d, J = 8.0 Hz, 1H), 8.43 (d, J = 8.4 Hz, 1H), 8.80 (m, 2H) [quinaldinate signals].

*Preparation of [Zn(quin)₂(4-Pyridone)] (**6**).* A teflon container was loaded with [Zn(quin)₂(CH₃OH)₂] (50 mg, 0.11 mmol), 4-hydroxypyridine (110 mg, w = 0.95, 1.10 mmol), triethylamine (250 mg, 2.47 mmol), and dried acetonitrile (10 mL). The container was closed and inserted into a steel autoclave. The autoclave was heated for 24 h at 105 °C. The reaction vessel was then allowed to cool slowly to room temperature. Block-like crystals of **6** of a gold yellow color were collected by filtration. The mass of the dried product was 32 mg. Yield: 0.063 mmol, 57%. Found C, 59.01; H, 3.70; N, 8.39%. $C_{25}H_{17}N_3O_5Zn$ requires C, 59.48; H, 3.39; N, 8.32%. IR (ATR, cm^{-1}): 3077w, 1644vvs [ν_{asym}(COO)], 1603vvs, 1570m, 1530vvs, 1462vs, 1367vvs [ν_{sym}(COO)], 1264m, 1201s, 1177s, 1151m, 1115w, 1049s, 1017vvs [ν(C–O), 4-pyridone], 965m, 900m, 872w, 848vvs, 804vvs, 775vvs, 739m, 639s, 606s. ^{1}H NMR (DMSO) δ/ppm: 6.13 (d, J = 6.6 Hz, 2H), 7.61 (d, J = 5.0 Hz, 2H) [4-pyridone signals]; 7.77 (m, 2H), 7.87 (m, 2H), 8.17 (d, J = 8.1 Hz, 2H), 8.35 (d, J = 8.4 Hz, 2H), 8.47 (d, J = 8.2 Hz, 2H), 8.76 (d, J = 8.3 Hz, 2H) [quinaldinate signals].

*Preparation of [Zn(quin)₂(4-Hmpy)₂] (**7**).* A teflon container was loaded with [Zn(quin)₂(CH₃OH)₂] (50 mg, 0.11 mmol), 4-hydroxymethylpyridine (110 mg, 1.00 mmol), and dried acetonitrile (15 mL). The container was closed and inserted into a steel autoclave. The autoclave was heated for 24 h at 105 °C. The reaction vessel was then allowed to cool slowly to room temperature. Large block-like crystals of **7** of a gold yellow color were collected by filtration. The mass of the dried product was 36 mg. Yield: 0.057 mmol, 52%. Found C, 61.21; H, 4.06; N, 8.89%. $C_{32}H_{26}N_4O_6Zn$ requires C, 61.20; H, 4.17; N, 8.92%. IR (ATR, cm^{-1}): 3287m [ν(O–H), 4-Hmpy], 3076w, 2832w, 1632vvs [ν_{asym}(COO)], 1602s, 1565m, 1555s, 1505m, 1458s, 1418s, 1367vvs [ν_{sym}(COO)], 1343vs, 1318s, 1270m, 1235w, 1215s, 1204m, 1176s, 1148m, 1114w, 1095s, 1067vs, 1022w, 1010vs [ν(C–O), 4-Hmpy], 996m, 963s, 897vs, 854vs, 802vvs, 779vvs, 720m, 687s, 665w, 637vs. ^{1}H NMR (DMSO) δ/ppm: 4.53 (d, J = 5.6 Hz, 2H), 5.40 (t, J = 5.7 Hz, 1H), 7.31 (d, J = 5.8 Hz, 2H), 8.41 (d, J = 5.9 Hz, 2H) [4-hydroxymethylpyridine signals]; 7.82 (m, 1H), 8.00 (m, 1H), 8.20 (d, J = 8.1 Hz, 1H), 8.41 (d, J = 8.4 Hz, 1H), 8.80 (m, 2H) [quinaldinate signals].

2.3. X-ray Structure Determinations

Each crystal was mounted on the tip of a glass fiber with a small amount of silicon grease and transferred to a goniometer head. Data were collected on an Agilent SuperNova diffractometer (Agilent Technologies XRD Products, Oxfordshire, UK), using graphite-monochromatized Mo-Kα radiation at 150 K. Data reduction and integration were performed with the software package CrysAlis PRO [38]. Corrections for the absorption (multi-scan) were made in all cases. The coordinates of the majority of non-hydrogen atoms were found via direct methods using the structure solution program SHELXS [39]. The positions of the remaining non-hydrogen atoms were located by use of a combination of least-squares refinement and difference Fourier maps in the SHELXL-97 program [40]. The positions of NH, NH₂, and OH hydrogen atoms in **3–7** were unambiguously located from the residual electron density maps. All other hydrogen atoms were placed in geometrically calculated positions and refined using a riding model. A summary of crystal data and refinement parameters for **1–7** is given in Table 1. Figures depicting the structures were prepared by ORTEP3 [41], Mercury [42], and CrystalMaker [43].

Table 1. Crystallographic data for 1–7.

	1	2	3	4	5	6	7
Empirical formula	$C_{30}H_{22}N_4O_4Zn$	$C_{34}H_{30}N_4O_4Zn$	$C_{36}H_{30}N_8O_6Zn$	$C_{30}H_{22}N_4O_6Zn$	$C_{32}H_{26}N_4O_6Zn$	$C_{25}H_{17}N_3O_5Zn$	$C_{32}H_{26}N_4O_6Zn$
Formula weight	567.89	623.99	736.05	599.89	627.94	504.79	627.94
Crystal system	monoclinic	triclinic	monoclinic	monoclinic	triclinic	monoclinic	monoclinic
Space group	$P\,2_1/c$	$P\,\bar{1}$	$P\,2_1/n$	$P\,2_1/n$	$P\,\bar{1}$	$P\,2_1/n$	$P\,2_1/n$
T [K]	150(2)	150(2)	150(2)	150(2)	150(2)	150(2)	150(2)
a [Å]	8.8768(11)	8.0866(4)	8.7982(4)	9.8119(5)	7.1034(4)	11.2818(4)	8.8653(6)
b [Å]	9.5542(17)	9.2741(3)	14.0204(5)	9.6039(4)	9.6254(5)	15.0187(6)	17.9746(10)
c [Å]	15.592(2)	9.9121(4)	14.3149(7)	13.8912(9)	10.7604(6)	13.3256(6)	9.6797(6)
α [°]	90	96.315(3)	90	90	97.779(5)	90	90
β [°]	105.972(13)	96.506(4)	105.608(5)	102.700(6)	109.171(5)	101.321(4)	116.307(8)
γ [°]	90	100.997(3)	90	90	99.391(4)	90	90
V [Å³]	1271.3(3)	718.39(5)	1700.69(13)	1276.98(12)	671.37(6)	2213.93(15)	1382.71(15)
D_{calcd} [g/cm³]	1.484	1.442	1.437	1.56	1.553	1.153	1.508
Z	2	1	2	2	1	4	2
λ [Å]	0.71073	0.71073	0.71073	0.71073	0.71073	0.71073	0.71073
μ [mm⁻¹]	1.011	0.902	0.781	1.017	0.971	1.153	0.943
Collected reflections	6568	6544	10541	7407	5739	12358	6568
Unique reflections, R_{int}	2899, 0.0839	3309, 0.0287	3907, 0.0490	2886, 0.0349	3077, 0.0304	5082, 0.0320	3179, 0.0267
Observed reflections	1931	3054	2849	2245	2788	4254	2621
$R1$ [1] $(I > 2\sigma(I))$	0.052	0.0379	0.0409	0.034	0.0364	0.0325	0.0302
$wR2$ [2] (all data)	0.1206	0.097	0.0915	0.075	0.0878	0.0801	0.0748

[1] $R1 = \sum||F_o|-|F_c||/\sum|F_o|$. [2] $wR2 = [\sum[w(F_o^2-F_c^2)^2]/\sum[w(F_o^2)^2]]^{1/2}$.

3. Results and Discussion

3.1. Synthetic Aspects

A straightforward synthesis of heteroleptic Zn(II) complexes with quinaldinate and a secondary pyridine ligand was based on a facile substitution of labile methanol ligands in [Zn(quin)$_2$(CH$_3$OH)$_2$]. The {Zn(quin)$_2$} core with quinaldinates bound in a chelating manner was not expected to undergo any significant structural changes. With the exception of the hydroxypyridine ligands, the proposed strategy resulted in novel complexes in reasonable yields. An excess of pyridine ligand was required to avoid the coordination of water to the zinc ion. The amount of water remaining in the solvent after drying was sufficient for the formation of an aqua complex. The reaction outcomes of the 3- or 4-hydroxypyridine systems were not as anticipated. Hydroxypyridines are prone to enol-ketonic tautomerism which strongly affects their nature, including their ability to function as ligands [44,45]. For 4-hydroxypyridine, the keto form is a dominant one in the solid state. The product isolated from the 4-Py-OH reaction mixture is [Zn(quin)$_2$(4-Pyridone)] (**6**), a complex with pyridone. It is to be noted that for its isolation the reaction mixture had to contain a stoichiometric amount of triethylamine. In the absence of base, no complex of zinc(II) with any form of 4-Py-OH was obtained. Following literature examples [46,47], the amine should assist in deprotonation of hydroxypyridine with the resulting anionic species engaging both donor sites in coordination. In the case of 3-Py-OH, the ligand in the solid state is known to consist of a mixture of both the neutral and the bipolar forms connected by strong intermolecular hydrogen bonds [48]. The formation of zwitterionic species is associated mainly with the acid-base properties of 3-Py-OH. Upon the protonation of nitrogen, the ligand's tendency towards ligation of metal ions is highly diminished. A negligible yield of [Zn(quin)$_2$(3-Py-OH)$_2$] (**4**) speaks in favor of the zwitterion being a major contributor also in the reaction system described in this paper.

3.2. Description of Structures

The title zinc(II) quinaldinate complexes with pyridine ligands are neutral. The first coordination sphere of zinc(II) ion consists in all but a 4-pyridone complex **6** of two quinaldinates and two pyridine-based ligands in an all trans arrangement. The N$_4$O$_2$ donor set renders a distorted octahedral environment around the zinc(II) ion. Because of the similarity of complexes, their molecular structures will be described together. They crystallize in centrosymmetric space groups with zinc ions located on the inversion centers. The coordination molecules possess an inherent crystallographic symmetry which places them in C$_i$ point group. In each case, the asymmetric unit consists of a half of a complex molecule. The Oak Ridge Thermal Ellipsoid Plot (ORTEP) drawings of [Zn(quin)$_2$(Py)$_2$] (**1**), [Zn(quin)$_2$(Nia)$_2$]·2CH$_3$CN (**3**), and [Zn(quin)$_2$(3-Py-OH)$_2$] (**4**) are shown in Figures 1–3, whereas the drawings of [Zn(quin)$_2$(3,5-Lut)$_2$] (**2**), [Zn(quin)$_2$(3-Hmpy)$_2$] (**5**), and [Zn(quin)$_2$(4-Hmpy)$_2$] (**7**) are given in the Supplementary Materials.

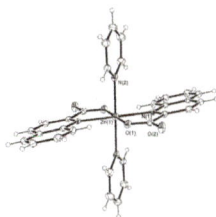

Figure 1. The ORTEP drawing of [Zn(quin)$_2$(Py)$_2$] (**1**). Atoms are represented by displacement ellipsoids at the 30% probability level. Hydrogen atoms are shown as spheres of arbitrary radii. Selected bond lengths [Å] and angles [°]: Zn(1)–N(1) = 2.225(13), Zn(1)–O(1) = 2.0249(19), Zn(1)–N(2) = 2.214(2), O(1)–Zn(1)–N(1) = 79.52(9), O(1)–Zn(1)–N(2) = 90.81(8), N(1)–Zn(1)–N(2) = 93.76(10).

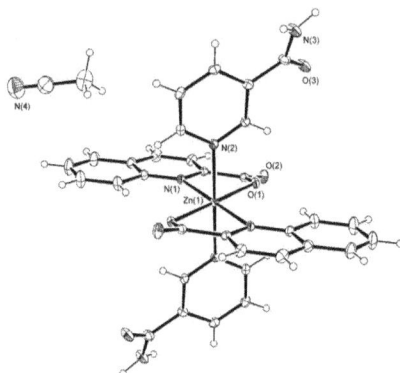

Figure 2. The ORTEP drawing of [Zn(quin)$_2$(Nia)$_2$]·2CH$_3$CN (**3**). Atoms are represented by displacement ellipsoids at the 30% probability level. Hydrogen atoms are shown as spheres of arbitrary radii. Selected bond lengths [Å] and angles [°]: Zn(1)–N(1) = 2.2248(19), Zn(1)–O(1) = 2.0519(14), Zn(1)–N(2) = 2.1861(17), O(1)–Zn(1)–N(1) = 78.21(6), O(1)–Zn(1)–N(2) = 91.74(6), N(1)–Zn(1)–N(2) = 92.92(7).

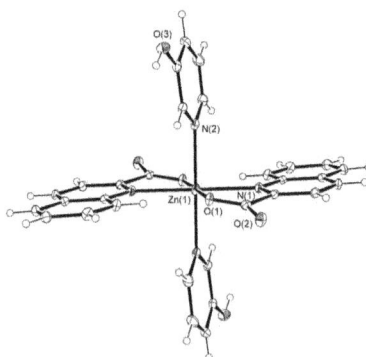

Figure 3. The ORTEP drawing of [Zn(quin)$_2$(3-Py-OH)$_2$] (**4**). Atoms are represented by displacement ellipsoids at the 30% probability level. Hydrogen atoms are shown as spheres of arbitrary radii. Selected bond lengths [Å] and angles [°]: Zn(1)–N(1) = 2.2298(15), Zn(1)–O(1) = 2.0391(13), Zn(1)–N(2) = 2.2008(17), O(1)–Zn(1)–N(1) = 78.86(5), O(1)–Zn(1)–N(2) = 91.36(6), N(1)–Zn(1)–N(2) = 90.44(6).

The relevant bonding parameters are summed up in Table 2. Pyridine-based ligands are coordinated via their nitrogen atoms. The pendant OH or CH$_2$OH groups of 3-Py-OH, 3-Hmpy, and 4-Hmpy ligands did not interact with zinc(II) ion. The zinc-to-pyridine ligand bond lengths span a somewhat wide range, 2.1861(17)–2.2420(14) Å, with the shortest ones occurring in the nicotinamide complex **3** and the 3-Hmpy complex **5** while the longest one occurs in the 4-Hmpy complex **7**. The quinaldinate has adopted its most common binding mode, the *N,O*-chelating coordination with the zinc-to-nitrogen bond lengths in the 2.2214(14)–2.2462(17) Å range and the zinc-to-oxygen bond lengths in the 2.0208(11)–2.0519(14) Å range. The series displays a pronounced similarity in the binding parameters. With a chelating quinaldinate coordination to zinc(II), a stable five-membered chelate ring is formed with bite angles in the 77.94(6)–79.52(9)° range. The chelate ring is not planar.

Table 2. Relevant bonding parameters [Å, °] of novel zinc(II) quinaldinate complexes.

Parameter	1	2	3	4	5	6	7
Zn–quin⁻							
Zn–N	2.225(3)	2.2277(16)	2.2248(19)	2.2298(15)	2.2462(17)	2.1425(15) and 2.1496(15)	2.2214(14)
Zn–O	2.0249(19)	2.0147(14)	2.0519(14)	2.0391(13)	2.0500(13)	1.9952(13) and 2.0333(13)	2.0208(11)
bite angle	79.52(9)	78.66(6)	78.21(6)	78.86(5)	77.94(6)	79.85(6) and 79.23(5)	78.06(5)
torsion angle [1]	6.9(4)	4.1(3)	11.4(3)	13.6(3)	7.6(3)	6.1(2) and 6.0(2)	2.9(2)
dihedral angle [2]	8.5(5)	5.5(3)	11.6(4)	13.7(3)	7.0(3)	7.1(3) and 7.4(3)	2.7(3)
C–O	1.230(3),	1.225(3),	1.238(3),	1.241(2),	1.235(2),	1.228(2), 1.282(2)	1.235(2),
	1.269(4)	1.273(2)	1.270(3)	1.263(2)	1.264(3)		1.271(2)
Zn–pyridine ligand							
L	Py	3,5-Lut	Nia	3-Py-OH	3-Hmpy	4-pyridone	4-Hmpy
donor atom	N	N	N	N	N	O	N
Zn–L	2.214(2)	2.2395(17)	2.1861(17)	2.2008(17)	2.1868(16)	1.9602(13)	2.2420(14)

[1] N–C(*ortho*)–C(carboxylate)–O(coordinated carboxylate oxygen). It gives the deviation from the planarity of the five-numbered chelate ring. [2] Defined as a dihedral angle between the plane of the quinaldinate ring and the carboxylate group.

Its non-planarity may be described by a torsion angle involving four of its atoms. The largest value, i.e., 13.6(3)°, calculated for 3-hydroxypyridine complex [Zn(quin)$_2$(3-Py-OH)$_2$] (**4**), corresponds to the least planar ring. As a consequence, the carboxylate functional group is not coplanar with the quinoline plane. Its non-coordinated or *free* oxygen atom points out from the aromatic plane. The disruption of planarity can be measured by a dihedral angle formed between the quinoline plane and the carboxylate plane. The largest value was observed again in the 3-Py-OH complex **4**, whereas the 4-Hmpy complex **7** shows the smallest deviation from planarity. A twist of carboxylate with respect to the quinoline ring was noted previously for a related compound, [Zn(quin)$_2$(1-methylimidazole)$_2$] [49]. The distortion was attributed to the involvement of *free* carboxylate oxygen in intermolecular interactions. As will be shown presently, the *free* carboxylate oxygen atoms in compounds **1–5** and **7** also participate in intermolecular interactions. For each complex in the homologous series, the extent of distortion was compared to the strength of intermolecular interactions. It turns out that obvious correlation exists only for [Zn(quin)$_2$(3-Py-OH)$_2$] (**4**) where the greatest deviation corresponds to the shortest and thus the strongest interaction, i.e., a hydrogen bond of the O–H⋯COO⁻ type. Contrary to expectations, [Zn(quin)$_2$(Py)$_2$] (**1**), which lacks strong intermolecular interactions, still displays a non-negligible distortion.

The 4-Pyridone complex [Zn(quin)$_2$(4-Pyridone)] (**6**) differs from other members of the series. Its ORTEP drawing is presented in Figure 4. The first coordination sphere of zinc(II) consists of two quinaldinates and a single 4-pyridone ligand, rendering the N$_2$O$_3$ donor set. The pyridone ligand has coordinated to zinc(II) through the phenoxide oxygen atom. The molecule possesses no intrinsic crystallographic symmetry. The pair of quinaldinate ions, bound in the usual *N,O*-chelating manner with bite angles of 79.23(5) and 79.85(6)°, is no longer nearly coplanar as in the case of the other six complexes. In the absence of the fourth ligand, the quinaldinates bend away from the equatorial plane. The value of a dihedral angle between the quinaldinates amounts to 52.65(4)°. Furthermore, the zinc-to-quinaldinate bonds in **6** are significantly shorter than in the other six complexes. The orientation of pyridone is skewed with respect to the Zn–O vector, an angle of 136.40(6)° is formed between the latter and the O–N pyridone axis. The H⋯Zn distance is just below the 3.2 Å cut-off, the limit in considering the pseudo agostic H⋯M (M = transition metal cation) bonds [50,51]. Nevertheless, in complex **6** it is more the ligand structure that holds the C–H bond close to the metal than the attraction of C–H to the metal ion.

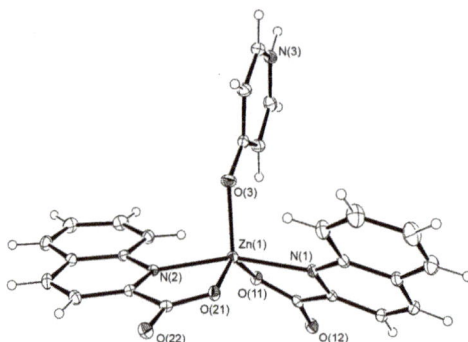

Figure 4. The ORTEP drawing of [Zn(quin)$_2$(4-Pyridone)] (**6**). Atoms are represented by displacement ellipsoids at the 30% probability level. Hydrogen atoms are shown as spheres of arbitrary radii. Selected bond lengths [Å] and angles [°]: Zn(1)–N(1) = 2.1425(15), Zn(1)–O(11) = 1.9952(13), Zn(1)–N(2) = 2.1496(15), Zn(1)–O(21) = 2.0333(13), Zn(1)–O(3) = 1.9602(13), O(11)–Zn(1)–N(1) = 79.85(6), O(21)–Zn(1)–N(2) = 79.23(5), O(21)–Zn(1)–O(11) = 138.17(5), N(2)–Zn(1)–N(1) = 161.26(6), O(3)–Zn(1)–N(1) = 100.45(6), O(3)–Zn(1)–O(11) = 115.49(6), O(3)–Zn(1)–N(2) = 96.84(6), O(3)–Zn(1)–O(21) = 106.13(5).

All known zinc(II) quinaldinate complexes possess quinaldinato ligands coordinated in the *N,O*-chelating fashion. Not only the composition but also the arrangement of ligands in three of these compounds is similar to compounds **1–5** and **7**: the mutual arrangement of ligands in [Zn(quin)$_2$(1-methylimidazole)$_2$] [49], [Zn(quin)$_2$(CH$_3$OH)$_2$] [35], and [Zn(quin)$_2$(DMSO)$_2$]·2DMSO [37] is *trans*. In contrast to the above examples, the arrangement of ligands in [Zn(quin)$_2$(Him)$_2$], a complex with imidazole, is *cis* [52]. As can be seen from Tables 2 and 3, the zinc-to-quinaldinate bond lengths in our compounds are of similar lengths to the ones shown in the cited examples. The remaining two literature examples, [Zn(Mepa)(quin)] (Mepa$^-$ = the anionic form of N-(2-mercaptoethyl)picolylamine) [53] and [Zn(quin)$_2$(H$_2$O)] [54], feature a five-numbered coordination environment of the metal. The arrangement of five donor atoms is intermediate between the square-pyramidal and trigonal-bipyramidal extremes. A degree of trigonality in five-coordinate systems is quantified by the geometric parameter, known as τ [55]. The value of 0.57 for [Zn(quin)$_2$(H$_2$O)] suggests a slightly greater distortion away from the tetragonal than for [Zn(quin)$_2$(4-Pyridone)] (**6**) with the τ value of 0.39. An appropriate description of the coordination polyhedron of the 4-pyridone complex is thus that of a distorted square pyramid with pyridone oxygen located at the apical site. In **6**, the zinc(II) ion lies ca. 0.5 Å above the basal plane, defined by the quinaldinate donor atoms. The similarity of bonding parameters of 4-pyridone complex **6** and aqua complex is obvious. Both display significantly shorter bonds than complexes with six donor atoms. This adds validity to the rule that with a smaller number of ligands, the bonds in the immediate vicinity of the metal ion are shorter.

Table 3. Relevant bonding parameters [Å, °] of zinc complexes with quinaldinate, literature data.

Complex	Zn–N	Zn–O	Zn–L	Ref.
[Zn(quin)$_2$(1-methylimidazole)$_2$]	2.244(2)	2.057(2)	2.144(2)	[49]
[Zn(quin)$_2$(CH$_3$OH)$_2$]	2.231(1)	2.003(1)	2.170(2)	[35]
cis-[Zn(quin)$_2$(Him)$_2$]	2.293(4)–2.418(4)	2.014(3)–2.052(3)	2.088(4)–2.135(4)	[52]
[Zn(Mepa)(quin)]	2.220(4)	2.015(2)	/^{2}	[53]
[Zn(quin)$_2$(H$_2$O)] ^{1}	2.144(4), 2.147(4)	2.017(4), 1.994(4)	1.986(3)	[54]

^{1} The complex has no inherent crystallographic symmetry. Two sets of zinc-to-quinaldinate bond distances are given thereby. ^{2} Not relevant to this study.

Surprisingly, in spite of the commercial availability of pyridine-based ligands, their zinc complexes with the exception of pyridine itself and nicotinamide are extremely rare. The scarce zinc(II) complexes with the enol form of 3-hydroxypyridine serving as a ligand invariably display coordination through a nitrogen atom. A single 3-Py-OH ligand in a penta-coordinated zinc(II) complex with the [Zn(C$_4$H$_8$NOS$_2$)$_2$(3-Py-OH)] composition (where C$_4$H$_8$NOS$_2$$^-$ denotes N-(2-hydroxyethyl)-N-methyldithiocarbamate) is bound at a distance of 2.0375(16) Å [56]. A significantly shorter bond than the one determined for [Zn(quin)$_2$(3-Py-OH)$_2$] (**4**) is yet another manifestation of the influence of the number of ligands over the bonding pattern (see above). The keto form of 3-hydroxypyridine can also coordinate to zinc(II). In [Zn$_2$(N$_3$)$_4$(3-Pyridone)]·2H$_2$O, an azide complex with a two-dimensional layer structure, 3-pyridone is coordinated via an oxygen atom to two metal centers with Zn–O distances of 2.106(4) Å [57]. Furthermore, the literature reports on a zinc complex with the anionic form of 3-Py-OH, i.e., a 3-pyridinolato ion. 3-pyridinolate whose negative charge is localized on oxygen links a zinc porphyrin moiety with a boron subphthalocyanine [46]. The anion is coordinated to zinc via its pyridine nitrogen at a distance of 2.155(3) Å. Similar compounds are known also for 4-pyridinolato ion [46,47]. The title compound [Zn(quin)$_2$(4-Pyridone)] (**6**) bears more resemblance to [ZnCl$_2$(4-Pyridone)$_2$] [58]. In the latter, the coordination environment of the metal ion is tetrahedral with 4-pyridone ligands coordinated via O atoms at distances of 1.947 Å, distances which are very similar to the one in [Zn(quin)$_2$(4-Pyridone)] (**6**). The latter compound was obtained from the aqueous solution of zinc chloride and 4-Py-OH. *catena*-[Zn$_2$(4-Pyridone)(N$_3$)$_4$]·H$_2$O exemplifies another binding mode of 4-pyridone: a μ_2-bridging coordination via oxygen with two distinctly different Zn–O distances, i.e., 2.1316(16) and 2.2106(15) Å [59]. It is pertinent to note that no complex of zinc(II) with the enol form of 4-Py-OH has been reported so far [34]. When binding to zinc(II), 3-hydroxymethylpyridine employs two coordination modes. A monodentate coordination via pyridine nitrogen was observed in a heptametallic zinc cluster with Zn–N distances of 2.042(3) Å [60], whereas a polymeric zinc complex with benzoate features a bidentate bridging coordination via both functional groups with Zn–N distance of 2.141(1) Å and Zn–O distance of 2.204(1) Å [61]. The corresponding bond in [Zn(quin)$_2$(3-Hmpy)$_2$] (**5**) is longer. No other than a monodentate binding mode through pyridine nitrogen has been observed so far for 4-hydroxymethylpyridine. It is to be noted that the zinc-to-nitrogen bond length [2.021(2) Å] in a tetranuclear zinc phosphate [62] is also significantly shorter than in our complex [Zn(quin)$_2$(4-Hmpy)$_2$] (**7**).

In contrast to other pyridine ligands used in this study, pyridine itself and 3,5-lutidine possess no functional groups that could participate in hydrogen-bonding interactions. Therefore only weak intermolecular interactions may be observed in the solid state structures of [Zn(quin)$_2$(Py)$_2$] (**1**) and [Zn(quin)$_2$(3,5-Lut)$_2$] (**2**). In the pyridine complex, C–H···O contacts, i.e., C(21)···O(2) [$-x, y + 0.5, -z + 0.5$] = 3.101(4) and C(22)···O(2) [$-x, y + 0.5, -z + 0.5$] = 3.106(5) Å, which occur between pyridine carbon atoms and a non-coordinated or *free* carboxylate oxygen link molecules into layers that are coplanar with the *bc* plane. The arrangement of complex molecules within layers is such that a $\pi\cdots\pi$ stacking interaction occurs between pairs of quinaldinate rings: a pyridine part of one quinaldinate makes a close approach to an arene part of another. With quinaldinate ligands two other types of $\pi\cdots\pi$ stacking interactions are possible: an interaction of two arene moieties or an interaction of two pyridine moieties. According to Janiak [63], there is a prevalent occurrence of pyridine···arene interactions. The relevant parameters of this interaction in **1** are: centroid–centroid distance = 3.593(2) Å, interplanar distance = 3.351(1) Å, dihedral angle = 1.5(1)°, and offset angle = 21.1° [63]. The connectivity pattern in the crystal structure of [Zn(quin)$_2$(3,5-Lut)$_2$] (**2**) is different. The C–H···O contacts are of similar length, i.e., C(2)···O(2) [$-x + 3, -y + 1, -z$] = 3.191(3) Å, but they occur between a quinaldinate C–H and a *free* carboxylate oxygen. They link molecules into infinite chains which are held together via various $\pi\cdots\pi$ stacking interactions occurring between aromatic rings and via T-shaped C–H···π interactions which occur between the 3,5-lutidine C–H and the arene part of quinaldinate (Table S2).

The solid state structures of other compounds are governed by stronger interactions. Following the general rule, all good hydrogen-bond donors and acceptors are engaged in hydrogen-bonding [64]. Hydrogen-bonding interactions are summed up in Table 4. For a complete list of weaker intermolecular interactions see Tables S1–S7. A well-known synthon [65] in supramolecular chemistry may be recognized in the structure of [Zn(quin)$_2$(Nia)$_2$]·2CH$_3$CN (**3**), i.e., an amide···amide homosynthon with the N–H···O(carbonyl group) distance of 2.975(3) Å (Figure 5).

Table 4. Stronger intermolecular interactions in **3–7**.

Compound	Functions, Engaged in an Interaction [1]	Contact [Å] [2,3]
3	NH$_2$···COO$^-$	N(3)···O(2) [$-x + 0.5, y + 0.5, -z + 0.5$] = 2.885(3)
4	NH$_2$···C=O(amide)	N(3)···O(3) [$-x + 1, -y, -z + 1$] = 2.975(3)
	OH···COO$^-$	O(3)···O(2) [$-x + 1.5, y - 0.5, -z + 1.5$] = 2.640(2)
5	OH···COO$^-$	O(3)···O(2) [$x + 1, y + 1, z$] = 2.697(2)
6	NH(pyridone)···COO$^-$	N(3)···O(22) [$x + 1.5, -y + 0.5, z - 0.5$] = 2.783(2)
	COO$^-$···COO^{-4}	O(11)···O(22) [$-x, -y + 1, -z + 1$] = 2.934(2)
7	OH···COO$^-$	O(3)···O(2) [$-x + 1, -y, -z + 2$] = 2.794(2)

[1] If not stated otherwise, COO$^-$ refers to a non-coordinated carboxylate oxygen atom. [2] For atom labels, see respective figures. [3] The distances may be compared to the corresponding sums of the van der Waals radii, 3.07 Å for N+O, and 3.04 Å for O+O [66]. [4] A short contact between a coordinated oxygen of one COO$^-$ and a non-coordinated oxygen from another. See text.

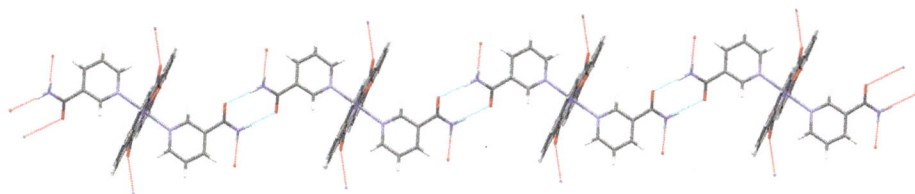

Figure 5. Section of a chain in [Zn(quin)$_2$(Nia)$_2$]·2CH$_3$CN (**3**), highlighting the amide⋯amide homosynthon.

With each complex molecule containing two equivalent nicotinamide ligands, molecules are linked into chains. The amide-dimer motif has been conveniently designated as R$_2^2$(8) [64]. The usually observed N–H⋯O contacts within the dimer are shorter, as exemplified by a co-crystal of 4-hydroxybenzamide/4,4'-bipyridine N,N'-dioxide with a distance of 2.892(3) Å [67]. The lengthening in **3** finds explanation in the involvement of the other NH$_2$ hydrogen atom in a hydrogen-bonding interaction with a non-coordinated carboxylate oxygen (shown in Figure 5 as a hanging dashed line) from an adjacent chain. Combination of both hydrogen-bonding motifs effectively links chains into layers which are coplanar with the [1 0 −1] plane (Figure 6). Each complex molecule within this layer forms altogether eight hydrogen bonds with its six closest neighbors. The layers interact via π⋯π stacking interactions with the shortest approach observed between the pyridine parts of the quinaldinate rings [centroid–centroid distance = 3.888(1) Å, interplanar distance = 3.4455(9) Å, dihedral angle = 0°, and offset angle = 27.6°]. According to Janiak [63], the latter are the least important among all three possible interactions for quinaldinate rings. The guest acetonitrile molecules are encapsulated between the layers.

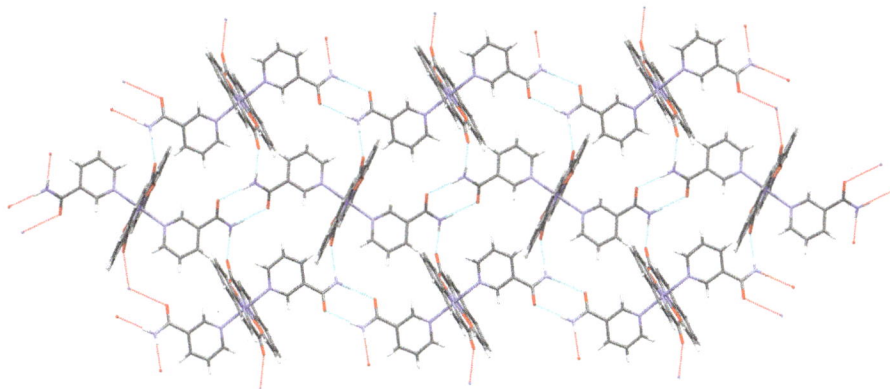

Figure 6. Section of a layer of [Zn(quin)$_2$(Nia)$_2$] molecules, linked via N–H⋯O bonds.

The presence of a hydroxyl moiety in complex molecules of [Zn(quin)$_2$(3-Py-OH)$_2$] (**4**), [Zn(quin)$_2$(3-Hmpy)$_2$] (**5**), and [Zn(quin)$_2$(4-Hmpy)$_2$] (**7**) determines their connectivity in the solid state. In all structures, the hydroxyl moiety is engaged in a strong hydrogen bond with a non-coordinated carboxylate oxygen with the shortest bond occurring in **4**, a complex with 3-hydroxypyridine (Table 4). In **4**, each molecule forms four O–H⋯O interactions with four neighboring molecules. As a result, layers that are coplanar with the [1 0 −1] plane are formed (Figure 7).

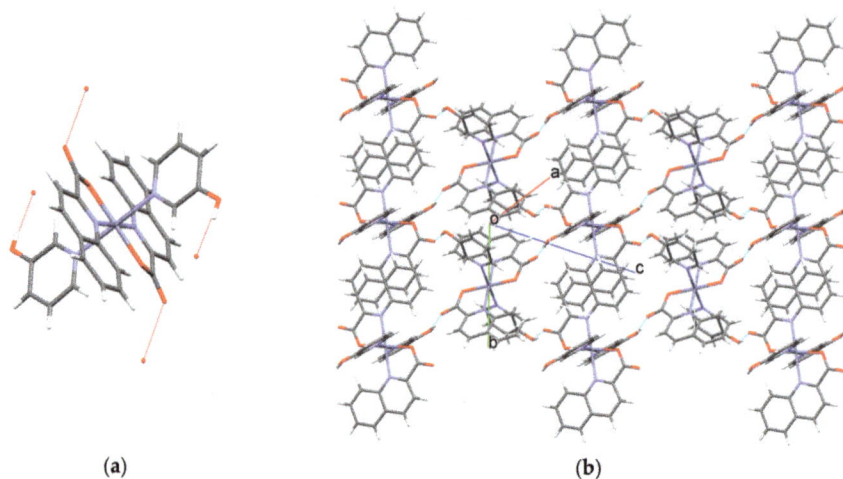

| (a) | (b) |

Figure 7. Hydrogen-bonding pattern in [Zn(quin)$_2$(3-Py-OH)$_2$] (**4**). (**a**) Each complex molecule is engaged in four O–H···COO$^-$ interactions. (**b**) Section of a layer.

A close proximity of molecules within layers allows π···π stacking interactions of quinaldinate rings and T-shaped C–H···π interactions between the quinaldinate C–H and the 3-Py-OH ring. The presence of a methylene group in 3-hydroxymethylpyridine makes the ligand more flexible when forming interactions with its surroundings. The O–H···O contact in [Zn(quin)$_2$(3-Hmpy)$_2$] (**5**) is slightly longer than in [Zn(quin)$_2$(3-Py-OH)$_2$] (**4**). The main difference between the pair is shown in the connectivity pattern. In the case of 3-Hmpy, each complex molecule forms four hydrogen bonds with two adjacent molecules producing an infinite chain pattern that extends along the 1 1 0 vector (Figure 8). A similar pattern is displayed by [Zn(quin)$_2$(4-Hmpy)$_2$] (**7**), a complex with 4-hydroxymethylpyridine with the longest O–H···O contact among all (Figure 9).

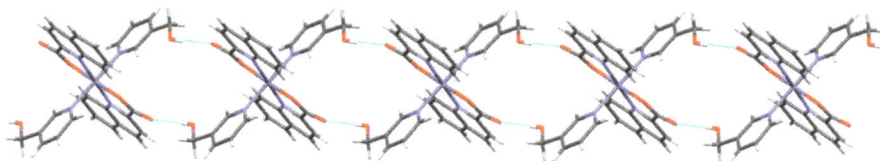

Figure 8. Hydrogen-bonding pattern in [Zn(quin)$_2$(3-Hmpy)$_2$] (**5**): a short section of a chain.

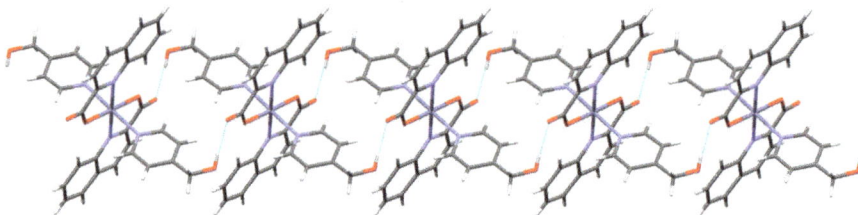

Figure 9. Hydrogen-bonding pattern in [Zn(quin)$_2$(4-Hmpy)$_2$] (**7**): a short section of a chain.

The packing of molecules in the structure of [Zn(quin)$_2$(3-Hmpy)$_2$] (**5**) is such that it allows $\pi\cdots\pi$ stacking interactions of 3-hydroxymethylpyridine rings [centroid–centroid distance = 4.297(1) Å, interplanar distance = 3.6011(8) Å, dihedral angle = 0°, and offset angle = 33.1°] and quinaldinate rings [an arene$\cdots$arene type, centroid–centroid distance = 3.614(1) Å, interplanar distance = 3.4007(9) Å, dihedral angle = 0°, and offset angle = 19.8°]. A closer inspection of the packing of molecules in the structure of [Zn(quin)$_2$(4-Hmpy)$_2$] (**7**) reveals no $\pi\cdots\pi$ stacking interactions of 4-hydroxymethylpyridine rings. Instead, there is a T-shaped C–H$\cdots\pi$ interaction which occurs between the quinaldinate C–H and the 4-Hmpy ring. The $\pi\cdots\pi$ stacking interactions between quinaldinate rings may also be observed [a pyridine$\cdots$arene type, centroid–centroid distance = 3.713(1) Å, interplanar distance = 3.3740(7) Å, dihedral angle = 1.4(1)°, and offset angle = 24.7°].

4-Pyridone of [Zn(quin)$_2$(4-Pyridone)] (**6**) is engaged via its NH group in hydrogen bonding interaction with COO$^-$ of another molecule. Complex molecules are thereby linked into infinite chains (Figure 10). The packing arrangement of [Zn(quin)$_2$(4-Pyridone)] molecules allows another type of interaction: a dipole$\cdots$dipole interaction between pairs of carbonyl moieties. Through the agency of the latter, dimeric entities are formed (illustrated in Figure 11). Within each dimer, all four carboxylates are involved. The carbonyl$\cdots$carbonyl interactions are well-known [68]. They appear as three geometric motifs, differing in the number of short C$\cdots$O contacts, i.e., distances that are significantly shorter than 3.6 Å. An antiparallel motif enables two short C$\cdots$O contacts, as exemplified by a tertiary squaramide, bis-3,4-(diethylamino)-cyclobutene-1,2-dione, with lengths of 2.879 Å [69]. On the other hand, a sheared parallel and a perpendicular arrangement are characterized by one short contact [68]. In compound **6**, the shortest C$\cdots$O contact [2.882(2) Å] occurs for a perpendicular arrangement of carbonyl moieties. The corresponding O$\cdots$O contacts [2.934(2) Å] are shown in Figure 11 as dashed lines. Within this dimer an antiparallel alignment of a pair of carbonyl moieties may also be observed. It is characterized by two slightly longer C$\cdots$O contacts [3.107(2) Å]. As a result of both the carbonyl$\cdots$carbonyl interactions and the NH$\cdots$COO$^-$ hydrogen bonds, an infinite layered structure is formed (Figure 12).

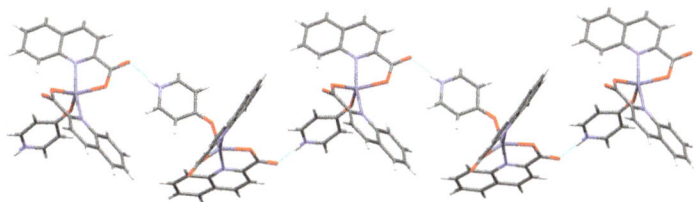

Figure 10. In [Zn(quin)$_2$(4-Pyridone)] (**6**), the N–H$\cdots$COO$^-$ interactions link molecules into chains.

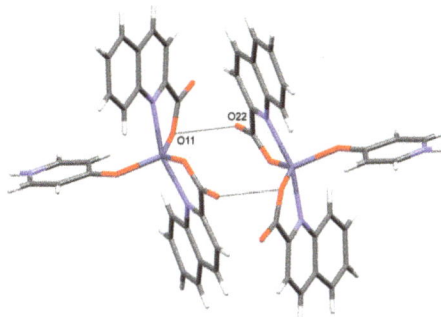

Figure 11. The carboxylate$\cdots$carboxylate interactions in **6**.

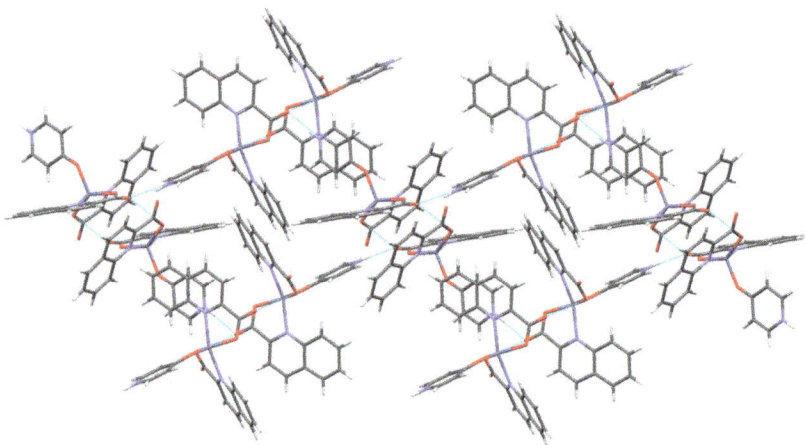

Figure 12. Linkage of [Zn(quin)₂(4-Pyridone)] molecules into layers in **6**.

3.3. Infrared Spectroscopy

The most relevant features of the infrared spectra of the title compounds concern the carbon–oxygen frequencies of the quinaldinato ligand and the characteristic absorptions of the pyridine-based ligands. The former are often used to diagnose the carboxylate binding mode [70,71]. In the case of a monodentate binding mode of the carboxylate, as in title compounds, a large difference in the positions of the ν_{asym}(COO) and ν_{sym}(COO) bands, known also as the splitting value Δ, may be observed. The title complexes display the splitting values Δ in the range 261–299 cm^{-1}. It is to be noted that values closer to 299 cm^{-1} were determined for [Zn(quin)₂(Py)₂] (**1**) and [Zn(quin)₂(3,5-Lut)₂] (**2**). Although in both compounds the *free* carboxylate oxygen participates in weak C–H···O interactions, the pair lacks stronger intermolecular interactions, i.e., hydrogen bonds. On the other hand, the spectra of the nicotinamide, 3-hydroxylmethylpyridine, 4-pyridone and 4-hydroxymethylpyridine complexes, where the carboxylate oxygen is engaged in hydrogen bonds, show values that are closer to 260 cm^{-1}. Our results are in line with the known influence of the strength of intermolecular interactions over the carboxylate splitting value Δ [52]. The positions of the ν_{asym}(COO) and ν_{sym}(COO) bands, i.e., 1654–1624 and 1368–1351 cm^{-1}, respectively, agree well with the data observed for related complexes, [Zn(quin)₂(1-methylimidazole)₂] [1642 and 1366 cm^{-1}] and [Zn(quin)₂(H₂O)] [1630 and 1390 cm^{-1}] [49,54].

Pyridine-based ligands, i.e., nicotinamide, 3-hydroxylmethylpyridine, and 4-hydroxymethylpyridine, have functional groups that can be unambiguously identified by infrared spectroscopy [72]. A strong band at 1692 cm^{-1} in the spectrum of [Zn(quin)₂(Nia)₂]·2CH₃CN (**3**) may be assigned to the stretching vibration of the C=O moiety of the amide, whereas the absorptions due to the asymmetric and symmetric stretching vibrations of the NH₂ moiety occur as broad bands at 3292 and 3147 cm^{-1}. Absorptions of the alcohol C–O stretching vibrations appear in the spectra of [Zn(quin)₂(3-Hmpy)₂] (**5**) and [Zn(quin)₂(4-Hmpy)₂] (**7**) in the 1095–1010 cm^{-1} region. Two or three very intense bands may be observed. Their positions are like the ones found in the spectra of pure ligands. The ν(O–H) absorption of the hydroxymethylpyridine ligands may be seen in the spectra of [Zn(quin)₂(3-Hmpy)₂] (**5**) and [Zn(quin)₂(4-Hmpy)₂] (**7**) as a sharp, medium intensity band at 3367 or 3287 cm^{-1}, respectively. In the spectra of pure hydroxymethylpyridines this band is broad. A change of the shape of the ν(O–H) band in the spectra of complexes, as compared to that of pure ligands, results from a more organized hydrogen-bonding pattern involving the OH moiety in the solid state structures of complexes **5** and **7**, i.e., hydrogen bonds of the OH···COO^{-} type and of moderate strength (Table 4). The absence of the ν(O–H) absorption in the spectrum of [Zn(quin)₂(4-Pyridone)] (**6**) confirms the keto form of the ligand.

*3.4. Thermal Analysis of [Zn(quin)₂(Py)₂] (**1**), [Zn(quin)₂(3,5-Lut)₂] (**2**), [Zn(quin)₂(3-Hmpy)₂] (**5**), [Zn(quin)₂(4-Hmpy)₂] (**7**), and [Zn(quin)₂(4-Pyridone)] (**6**)*

A pronounced similarity in the behavior of pyridine, 3,5-lutidine, 3- and 4-hydroxymethylpyridine complexes upon heating to 800 °C in argon was observed. In all, two well-resolved regions of major mass loss may be observed. Considering the higher volatility of pyridine-based ligands, the first step is due to their liberation from the zinc(II) coordination sphere, whereas the second one is governed by the pyrolysis of quinaldinate with the subsequent formation of zinc oxide. It is to be noted that the thermal stability of pyridine and 3,5-lutidine complexes is very similar. As will be shown, their decomposition starts at a 20–30 °C lower temperature than in other complexes. A narrower thermal stability window of the pair can be traced to the strength of intermolecular interactions in their solid state structures. The thermogravimetric (TG) curve of the pyridine complex, [Zn(quin)₂(Py)₂] (**1**), is shown in Figure 13, the TG curves of other compounds are given in Supplementary Materials. The first decomposition process of **1** starts at 155 °C and continues up to 230 °C. The accompanying mass loss is in excellent agreement with the release of pyridine ligands, calcd./found: 27.86%/27.90%. The residue {Zn(quin)₂} is stable until about 360 °C which marks the onset of the second decomposition process. The latter lasts up to ca. 570 °C and brings about a 49.12% decrease in mass. As shown by the differential scanning calorimetry (DSC) curve, both processes are endothermic. The final residual represents 28.97% of the initial mass. As confirmed by the powder X-ray diffraction, the resulting grey solid is a mixture of zinc oxide (calcd. 14.33%) with an amorphous material. The liberation of 3,5-lutidine ligands starts at 165 °C and continues up to 260 °C, calcd./found: 34.34%/34.31%. The second decomposition process occurs in the 330–500 °C interval with the 36.97% loss. The decomposition of the 4-Hmpy complex **7** starts at a slightly higher temperature, ca. 195 °C, and lasts up to 295 °C with the observed loss differing from the theoretical value for the release of 4-Hmpy molecules, calcd./found: 34.73%/29.16%. The first decomposition stage is immediately followed by the second which is completed at ca. 570 °C and is accompanied by the 39.91% mass loss. The TG curve of 3-Hmpy complex **5** reveals a small, yet a non-negligible reduction of mass, ca. 1–2%, in the 25–125 °C interval that is probably due to the solvent associated with the sample. The first major decomposition process in the 185–225 °C interval is followed by the second in the 230–290 °C interval with mass losses of 18.49% and 17.27%, respectively. Their sum is consistent with the liberation of 3-Hmpy molecules, calcd./found: 34.76%/35.76%. The final decomposition commences at 350 °C and lasts up to 570 °C. The inspection of thermal properties of [Zn(quin)₂(3-Hmpy)₂] (**5**) in the oxidizing atmosphere revealed a similar behavior with two major differences: (i) the onset of all processes is at lower temperatures, and (ii) the final decomposition stage is highly exothermic. A mass loss of 1.45% may be observed in the 80–125 °C interval, followed by a two-step dissociation of 3-Hmpy ligands in the 165–275 °C interval with the total loss amounting to 34.20%. The overall decomposition with the formation of zinc oxide starts at 290 °C and is completed by 490 °C, calcd./found: 12.96%/12.38%.

The TG curve of [Zn(quin)₂(4-Pyridone)] (**6**), obtained in argon atmosphere, differs from other studied compounds. It shows three distinct regions of mass loss. The first one is completed by 140 °C and accounts for ca. 7.5% of the initial mass. It may be ascribed to the solvent content, present in the sample. The latter could not be verified from the analytical data. The second one in the 215–335 °C interval is due to the 4-pyridone elimination, calcd./found: 18.84%/19.37%. The last one commences at 340 °C. The degradation of the compound continues at higher temperature without completion.

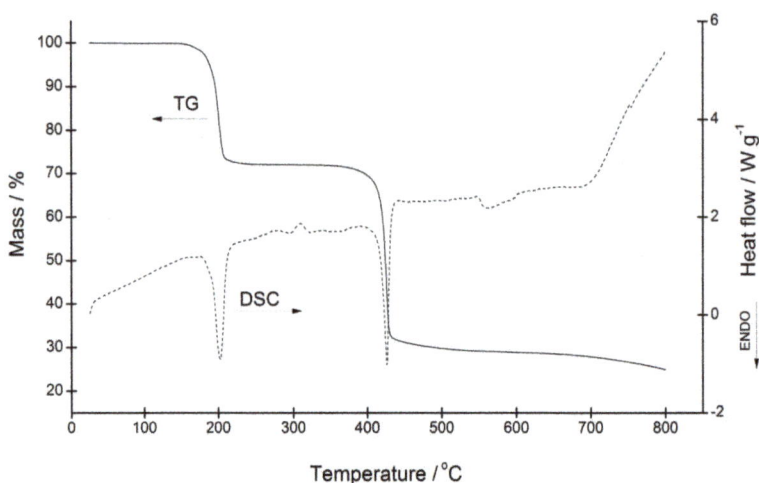

Figure 13. Thermogravimetric (TG) and differential scanning calorimetry (DSC) curves for [Zn(quin)$_2$(Py)$_2$] (**1**).

4. Conclusions

Structural chemistry of a series of heteroleptic zinc(II) complexes with quinaldinate and pyridine-based ligands is presented. Complexes displayed two stoichiometries: (i) [Zn(quin)$_2$L$_2$] (L = Py, 3,5-Lut, Nia, 3-Py-OH, 3-Hmpy and 4-Hmpy) with a nearly octahedral distribution of N$_4$O$_2$ donor sites, and [Zn(quin)$_2$(4-Pyridone)] with a square-pyramidal distribution of N$_2$O$_3$ donor sites. In all but a 4-hydroxypyridine compound, the pyridine-based ligand coordinated to the metal ion via its ring nitrogen atom. In the case of 4-Py-OH, its keto tautomeric form coordinated via an oxygen atom. Solid state structures of compounds displayed typical patterns of connectivity, determined by substituents of the pyridine-based ligands. For example, in [Zn(quin)$_2$(Nia)$_2$]·2CH$_3$CN (**3**) the predicted structure directing role of amide was realized through the formation of the amide···amide homosynthon. On the other hand, the OH moieties in the 3-Py-OH, 3-Hmpy and 4-Hmpy complexes invariably formed hydrogen bonds to the carboxylate.

For pyridine ligands containing the –CH$_2$OH functional group, new perspectives are opened with the introduction of a strong base to the system. Their anionic forms with the alkoxide oxygen atoms are known to engage both donor sites in coordination and could as such interfere with the structural integrity of the {Zn(quin)$_2$} core. This remains to be investigated in our future studies.

Supplementary Materials: The following are available online at www.mdpi.com/2073-4352/8/1/52/s1, ORTEP diagrams of **2** (Figure S1), **5** (Figure S2), and **7** (Figure S3). Packing diagrams of **1** (Figure S4), **2** (Figure S5), and **7** (Figure S6). Exhaustive lists of intermolecular interactions (Tables S1–S7). TG/DSC curves of **2** (Figure S7), **7** (Figure S8), **5** (Figure S9), **5** heated in the air (Figure S10), and **6** (Figure S11). Infrared spectra (Figures S12–S17). NMR spectra of DMSO-d_6 solutions of **5** (Figures S18 and S19) and **7** (Figures S20 and S21). CCDC- 1581372 (**1**), 1581373 (**2**), 1581374 (**3**), 1581375 (**4**), 1581376 (**5**), 1581377 (**6**), and 1581378 (**7**) contain the supplementary crystallographic data for this paper. These data can be obtained free of charge at http://www.ccdc.cam.ac.uk/conts/retrieving.html (or from the Cambridge Crystallographic Data Centre, 12 Union Road, Cambridge CB2 1EZ, UK; Fax: +44-1223-336033).

Acknowledgments: This work was supported by a grant from the Slovenian Ministry of Education, Science and Sport (Grant P1-0134). The author is grateful to Darko Dolenc, Janez Košmrlj, and Damijana Urankar for NMR measurements, to Romana Cerc Korošec for thermal analyses and to graduate students Valentina Brzuhalski and Daniela Mikic for testing the reproducibility of syntheses.

Author Contributions: Barbara Modec conceived and designed the experiments, carried out the syntheses, performed X-ray structure determinations and analyzed the results.

Conflicts of Interest: The author declares no conflict of interest.

References

1. Keilin, D.; Mann, T. Carbonic anhydrase. Purification and nature of the enzyme. *Biochem. J.* **1940**, *34*, 1163–1176. [CrossRef] [PubMed]
2. Christianson, D.W. The structural biology of zinc. *Adv. Prot. Chem.* **1991**, *42*, 281–335.
3. Aoki, S.; Kimura, E. Zinc hydrolases. In *Comprehensive Coordination Chemistry*; McCleverty, J.A., Meyer, T.J., Eds.; Elsevier: Amsterdam, Nederland, 2004; pp. 601–640.
4. Andreini, C.; Banci, L.; Bertini, I.; Rosato, A. Zinc through the three domains of life. *J. Proteome Res.* **2006**, *5*, 3173–3178. [CrossRef] [PubMed]
5. Vahrenkamp, H. Why does nature use zinc—A personal view. *Dalton Trans.* **2007**, 4751–4759. [CrossRef] [PubMed]
6. Sousa, S.F.; Lopes, A.B.; Fernandes, P.A.; Ramos, M.J. The zinc proteome: A tale of stability and functionality. *Dalton Trans.* **2009**, 7946–7956. [CrossRef] [PubMed]
7. Lipton, A.S.; Heck, R.W.; Ellis, P.D. Zinc solid-state NMR spectroscopy of human carbonic anhydrase: Implications for the enzymatic mechanism. *J. Am. Chem. Soc.* **2004**, *126*, 4735–4739. [CrossRef] [PubMed]
8. Auld, D.S. The ins and outs of biological zinc sites. *Biometals* **2009**, *22*, 141–148. [CrossRef] [PubMed]
9. Chasapis, C.T.; Loutsidou, A.C.; Spiliopoulou, C.A.; Stefanidou, M.E. Zinc and human health: An update. *Arch. Toxicol.* **2012**, *86*, 521–534. [CrossRef] [PubMed]
10. Tarushi, A.; Kljun, J.; Turel, I.; Pantazaki, A.A.; Psomas, G.; Kessissoglou, D.P. Zinc(II) complexes with the quinolone antibacterial drug flumequine: Structure, DNA- and albumin-binding. *New J. Chem.* **2013**, *37*, 342–355. [CrossRef]
11. Mjos, K.D.; Orvig, C. Metallodrugs in medicinal inorganic chemistry. *Chem. Rev.* **2014**, *114*, 4540–4563. [CrossRef] [PubMed]
12. Ni, L.; Wang, J.; Liu, C.; Fan, J.; Sun, Y.; Zhou, Z.; Diao, G. An asymmetric binuclear zinc(II) complex with mixed iminodiacetate and phenanthroline ligands: Synthesis, characterization, structural conversion and anticancer properties. *Inorg. Chem. Front.* **2016**, *3*, 959–968. [CrossRef]
13. Indoria, S.; Lobana, T.S.; Sood, H.; Arora, D.S.; Hundal, G.; Jasinski, J.P. Synthesis, spectroscopy, structures and antimicrobial activity of mixed-ligand zinc(II) complexes of 5-nitro-salicylaldehyde thiosemicarbazones. *New J. Chem.* **2016**, *40*, 3642–3653. [CrossRef]
14. Koleša-Dobravc, T.; Maejima, K.; Yoshikawa, Y.; Meden, A.; Yasui, H.; Perdih, F. Vanadium and zinc complexes of 5-cyanopicolinate and pyrazine derivatives: Synthesis, structural elucidation and in vitro insulin-mimetic activity study. *New. J. Chem.* **2017**, *41*, 735–746. [CrossRef]
15. McCall, K.; Huang, C.-C.; Fierke, C.A. Function and mechanism of zinc metalloenzymes. *J. Nutr.* **2000**, *130*, 1437S–1446S. [CrossRef] [PubMed]
16. Pearson, R.G. Hard and soft acids and bases. *J. Am. Chem. Soc.* **1963**, *85*, 3533–3539. [CrossRef]
17. Huheey, J.E.; Keller, E.A.; Keiter, R.L. *Inorganic Chemistry: Principles of Structure and Reactivity*, 4th ed.; Harper Collins College Publishers: New York, NY, USA, 1993; pp. 394–413, ISBN 0-06-042995-X.
18. Laitaoja, M.; Valjakka, J.; Janis, J. Zinc coordination spheres in protein structures. *Inorg. Chem.* **2013**, *52*, 10983–10991. [CrossRef] [PubMed]
19. Yao, S.; Flight, R.M.; Rouchka, E.C.; Moseley, H.N.B. A less-biased analysis of metalloproteins reveals novel zinc coordination geometries. *Proteins Struct. Funct. Genet.* **2015**, *83*, 1470–1487. [CrossRef] [PubMed]
20. Parkin, G. The bioinorganic chemistry of zinc: Synthetic analogues of zinc enzymes that feature tripodal ligands. *Chem. Commun.* **2000**, 1971–1985. [CrossRef]
21. Bosch, S.; Comba, P.; Gahan, L.R.; Schenk, G. Dinuclear zinc(II) complexes with hydrogen bond donors as structural and functional phosphatase models. *Inorg. Chem.* **2014**, *53*, 9036–9051. [CrossRef] [PubMed]
22. Linder, D.P.; Rodgers, K.R. Methanethiol binding strengths and deprotonation energies in Zn(II)-imidazole complexes from M05-2X and MP2 theories: Coordination number and geometry influences relevant to zinc enzymes. *J. Phys. Chem. B* **2015**, *119*, 12182–12192. [CrossRef] [PubMed]
23. Yoshikawa, Y.; Ueda, E.; Kawabe, K.; Miyake, H.; Takino, T.; Sakurai, H.; Kojima, Y. Development of new insulinomimetic zinc(II) picolinate complexes with a $Zn(N_2O_2)$ coordination mode: Structure characterization, in vitro, and in vivo studies. *J. Biol. Inorg. Chem.* **2002**, *7*, 68–73. [CrossRef] [PubMed]

24. Wang, Y.-T.; Fan, H.-H.; Wang, H.-Z.; Chen, X.-M. A solvothermally in situ generated mixed-ligand approach for NLO-active metal-organic framework materials. *Inorg. Chem.* **2005**, *44*, 4148–4150. [CrossRef] [PubMed]

25. Papatriantafyllopoulou, C.; Raptopoulou, C.P.; Terzis, A.; Janssens, J.F.; Manessi-Zoupa, E.; Perlepes, S.P.; Plakatouras, J.C. Assembly of a helical zinc(II) chain and a two-dimensional cadmium(II) coordination polymer using picolinate and sulfate anions as bridging ligands. *Polyhedron* **2007**, *26*, 4053–4064. [CrossRef]

26. Konidaris, K.F.; Polyzou, C.D.; Kostakis, G.E.; Tasiopoulos, A.J.; Roubeau, O.; Teat, S.J.; Manessi-Zoupa, E.; Powell, A.K.; Perlepes, S.P. Metal ion-assisted transformations of 2-pyridinealdoxime and hexafluorophosphate. *Dalton Trans.* **2012**, *41*, 2862–2865. [CrossRef] [PubMed]

27. Enthaler, S.; Wu, X.-F.; Weidauer, M.; Irran, E.; Döhlert, P. Exploring the coordination chemistry of 2-picolinic acid to zinc and application of the complexes in catalytic oxidation chemistry. *Inorg. Chem. Commun.* **2014**, *46*, 320–323. [CrossRef]

28. Mohammadnezhad, G.; Ghanbarpour, A.R.; Amini, M.M.; Ng, S.W. Bis(μ-quinoline-2-carboxylato)-$\kappa^3N,O:O;\kappa^3O:N,O$-bis[(acetato-$\kappa^2O,O'$)(methanol-$\kappa O$)lead(II)]. *Acta Cryst.* **2010**, *E66*, m963. [CrossRef] [PubMed]

29. Zurowska, B.; Mrozinski, J.; Ciunik, Z. Structure and magnetic properties of a copper(II) compound with *syn-anti* carboxylato- and linear Cu–Cl–Cu chloro-bridges. *Polyhedron* **2007**, *26*, 3085–3091. [CrossRef]

30. Zurowska, B.; Slepokura, K. Structure and magnetic properties of polynuclear copper(II) compounds with *syn-anti* carboxylato- and bromo-bridges. *Inorg. Chim. Acta* **2008**, *361*, 1213–1221. [CrossRef]

31. Chowdhury, A.D.; De, P.; Mobin, S.M.; Lahiri, G.K. Influence of nitrosyl coordination on the binding mode of quinaldate in selective ruthenium frameworks. Electronic structure and reactivity aspects. *RSC Adv.* **2012**, *2*, 3437–3446. [CrossRef]

32. Starynowicz, P. Structure of bis-μ-(2-quinolinecarboxylato-*O,O,O'*)bis[triaqua(2-quinolinecarboxylato-*N,O*)(2-quinolinecarboxylato-*O*)neodymium(III)] trihydrate. *Acta Cryst.* **1990**, *C46*, 2068–2070. [CrossRef]

33. Groom, C.R.; Bruno, I.J.; Lightfoot, M.P.; Ward, S.C. The Cambridge Structural Database. *Acta Cryst.* **2016**, *B72*, 171–179. [CrossRef] [PubMed]

34. Coxall, R.A.; Harris, S.G.; Henderson, D.K.; Parsons, S.; Tasker, P.A.; Winpenny, R.E.P. Inter-ligand reactions: In situ formation of new polydentate ligands. *J. Chem. Soc. Dalton Trans.* **2000**, 2349–2356. [CrossRef]

35. Yue, Z.-Y.; Cheng, C.; Gao, P.; Yan, P.-F. *Trans*-Bis(methanol-κO)bis(2-quinoline-carboxylato-κ^2N,O)zinc(II). *Acta Cryst.* **2004**, *E60*, m82–m84.

36. Williams, D.B.G.; Lawton, M. Drying of organic solvents: Quantitative evaluation of the efficiency of several desiccants. *J. Org. Chem.* **2010**, *751*, 8351–8354. [CrossRef] [PubMed]

37. Zhang, W.Z.; Shuang, M.; Zhu, M.C.; Lei, L.; Gao, E.J. Synthesis, crystal structure, and photoluminescence of zinc complex [Zn(Qina)$_2$(DMSO)$_2$]·2DMSO. *Russ. J. Coord. Chem.* **2009**, *35*, 874–879. [CrossRef]

38. *CrysAlis PRO*, Version 1.171.35.11; Agilent Technologies: Yarnton, Oxfordshire, UK, 2011.

39. Sheldrick, G.M. A short history of SHELX. *Acta Cryst.* **2008**, *A64*, 112–122. [CrossRef] [PubMed]

40. Sheldrick, G.M. Crystal structure refinement with SHELXL. *Acta Cryst.* **2015**, *C71*, 3–8.

41. Farrugia, J.J. ORTEP-3 for Windows—A version of ORTEP-III with a graphical user interface (GUI). *J. Appl. Crystallogr.* **1997**, *30*, 565. [CrossRef]

42. Macrae, C.F.; Edgington, P.R.; McCabe, P.; Pidcock, E.; Shields, G.P.; Taylor, R.; Towler, M.; van de Streek, J. Mercury: Visualization and analysis of crystal structures. *J. Appl. Cryst.* **2006**, *39*, 453–457. [CrossRef]

43. *CrystalMaker for Windows*, Version 2.6.1; CrystalMaker Software, Ltd.: Oxfordshire, UK, 2007.

44. Katritzky, A.R.; Lagowski, J.M. Prototropic tautomerism of heteroaromatic compounds: I. General discussion and methods of study. *Adv. Heterocycl. Chem.* **1963**, *1*, 311–338.

45. Rawson, J.M.; Winpenny, R.E.P. The coordination chemistry of 2-pyridone and its derivatives. *Coord. Chem. Rev.* **1995**, *139*, 313–374. [CrossRef]

46. Xu, H.; Ng, D.K.P. Construction of subphthalocyanine-porphyrin and subphthalocyanine heterodyads through axial coordination. *Inorg. Chem.* **2008**, *47*, 7921–7927. [CrossRef] [PubMed]

47. Choi, M.T.M.; Choi, C.-F.; Ng, D.K.P. Assembling tetrapyrole derivatives through axial coordination. *Tetrahedron* **2004**, *60*, 6889–6894. [CrossRef]

48. Koval'chukova, O.V.; Strashnova, S.B.; Zaitsev, B.E.; Vovk, T.V. Synthesis and physicochemical properties of some transition metal complexes with 3-hydroxypyridine. *Russ. J. Coord. Chem.* **2002**, *28*, 767–770. [CrossRef]

49. Zevaco, T.A.; Görls, H.; Dinjus, E. Synthesis, spectral and structural characterization of zinc carboxylate [Zn(2-quinolinecarboxylato)$_2$(1-methylimidazole)$_2$]. *Inorg. Chim. Acta* **1998**, *269*, 283–286. [CrossRef]

50. Braga, D.; Grepioni, F.; Tedesco, E.; Biradha, K.; Desiraju, G.R. Hydrogen bonding in organometallic crystals. 6. X–H···M hydrogen bonds and M···(H–X) pseudo-agostic bonds. *Organometallics* **1997**, *16*, 1846–1856. [CrossRef]

51. Brookhart, M.; Green, M.L.H.; Parkin, G. Agostic interactions in transition metal compounds. *Proc. Natl. Acad. Sci. USA* **2007**, *104*, 6908–6914. [CrossRef] [PubMed]

52. Dobrzynska, D.; Lis, T.; Jerzykiewicz, L.B. 3D supramolecular network constructed by intermolecular interactions in mixed ligand complex of zinc. *Inorg. Chem. Commun.* **2005**, *8*, 1090–1093. [CrossRef]

53. Brand, U.; Vahrenkamp, H. A new tridentate *N,N,S* ligand and its zinc complexes. *Inorg. Chem.* **1995**, *34*, 3285–3293. [CrossRef]

54. Okabe, N.; Muranishi, Y. Aquabis(quinoline-2-carboxylato-$\kappa^2 N, O$)zinc(II). *Acta Cryst.* **2003**, *E59*, m244–m246. [CrossRef]

55. Addison, A.W.; Rao, T.N.; Redijk, J.; van Rijn, J.; Verschoor, G.C. Synthesis, structure, and spectroscopic properties of copper(II) compounds containing nitrogen-sulphur donor ligands: The crystal and molecular structure of aqua[1,7-bis(*N*-methylbenzimidazol-2′-yl)-2,6-dithiaheptane]copper(II) perchlorate (τ is calculated using the equation $(\beta-\alpha)/60$ where α and β stand for basal angles. Its value is 0 for a perfect square pyramid, and it becomes unity for a trigonal bipyramid). *J. Chem. Soc. Dalton Trans.* **1984**, 1349–1356. [CrossRef]

56. Jotani, M.M.; Arman, H.D.; Poplaukhin, P.; Tiekink, E.R.T. Bis(*N,N*-diethyldithiocarbamato-$\kappa^2 S, S'$)(3-hydroxy pyridine-κN)zinc and bis[*N*-(2-hydroxyethyl)-*N*-methyldithiocarbamato-$\kappa^2 S, S'$](3-hydroxypyridine-κN)zinc: Crystal structures and Hirshfeld surface analysis. *Acta Cryst.* **2016**, *E72*, 1700–1709. [CrossRef] [PubMed]

57. Goher, M.A.S.; Sodin, B.; Bitschnau, B.; Fuchs, E.C.; Mautner, F.A. Ladders, rings and cubes as structural motifs in three new zinc(II) azide complexes: Synthesis, spectral and structural characterization. *Polyhedron* **2008**, *27*, 1423–1431. [CrossRef]

58. Masse, R.; Fur, Y.L. Crystal structure of bis(4-pyridone)dichlorozinc(II), $ZnCl_2(C_5H_5NO)_2$. *Zeitschrift für Kristallographie New Cryst. Struct.* **1998**, *213*, 114.

59. Mautner, F.A.; Berger, C.; Gspan, C.; Sudy, B.; Fisher, R.C.; Massoud, S.S. Pyridyl and triazole ligands directing the assembling of zinc(II) into coordination polymers with different dimensionality through azides. *Polyhedron* **2017**, *130*, 136–144. [CrossRef]

60. Feazell, R.P.; Carson, C.E.; Klausmeyer, K.K. Synthesis of a functionalized monomer of the rare $M_7O_2{}^{10+}$ double tetrahedron Zn cluster. *Inorg. Chem. Commun.* **2007**, *10*, 873–875. [CrossRef]

61. Zelenak, V.; Cisarova, I.; Sabo, M.; Llewellyn, P.; Gyoryova, K. A Zn(II)-coordination polymer formed by benzoate and 3-pyridinemethanol ligands: Synthesis, spectroscopic properties, crystal structure and kinetics of thermal decomposition. *J. Coord. Chem.* **2004**, *57*, 87–96.

62. Murugavel, R.; Kuppuswamy, S.; Boomishankar, R.; Steiner, A. Hierarchical structures built from a molecular zinc phosphate core. *Angew. Chem. Int. Ed.* **2006**, *45*, 5536–5540. [CrossRef] [PubMed]

63. Janiak, C. A critical account on $\pi-\pi$ stacking in metal complexes with aromatic nitrogen-containing ligands. *J. Chem. Soc. Dalton Trans.* **2000**, *21*, 3885–3896. [CrossRef]

64. Etter, M.C. Encoding and decoding hydrogen-bond patterns of organic compounds. *Acc. Chem. Res.* **1990**, *23*, 120–126. [CrossRef]

65. Aakeröy, C.B.; Scott, B.M.T.; Smith, M.M.; Urbina, J.F.; Desper, J. Establishing amide···amide reliability and synthon transferability in the supramolecular assembly of metal-containing one-dimensional architectures. *Inorg. Chem.* **2009**, *48*, 4052–4061. [CrossRef] [PubMed]

66. Douglas, B.B.; McDaniel, D.H.; Alexander, J.J. *Concepts and Models of Inorganic Chemistry*, 3rd ed.; Wiley: New York, NY, USA, 1994; ISBN 978-0-471-62978-8.

67. Babu, N.J.; Reddy, L.S.; Nangia, A. Amide-*N*-oxide heterosynthon and amide dimer homosynthon in cocrystals of carboxamide drugs and pyridine *N*-oxides. *Mol. Pharm.* **2007**, *4*, 417–434. [CrossRef] [PubMed]

68. Allen, F.H.; Baalham, C.A.; Lommerse, J.P.M.; Raithby, P.R. Carbonyl-carbonyl interactions can be competitive with hydrogen bonds. *Acta Cryst.* **1998**, *B54*, 320–329. [CrossRef]

69. Prohens, R.; Portell, A.; Font-Bardia, M.; Bauza, A.; Frontera, A. Experimental and theoretical study of weak intermolecular interactions in crystalline tertiary squaramides. *CrystEngComm* **2016**, *18*, 6437–6443. [CrossRef]

70. Nakamoto, K. *Infrared and Raman Spectra of Inorganic and Coordination Compounds. Part B: Applications in Coordination, Organometallic, and Bioinorganic Chemistry*, 5th ed.; John Wiley & Sons, Inc.: New York, NY, USA, 1997; pp. 23–30, 57–62, ISBN 0-471-16392-9.

71. Deacon, G.B.; Phillips, R.J. Relationships between the carbon–oxygen stretching frequencies of carboxylato complexes and the type of carboxylate coordination. *Coord. Chem. Rev.* **1980**, *33*, 227–250. [CrossRef]

72. Colthup, N.B.; Daly, L.H.; Wiberley, S.E. *Introduction to Infrared and Raman Spectroscopy*; Academic Press International Edition: New York, NY, USA, 1964; pp. 263, 274.

![crystals logo] *crystals*

MDPI

Article

Room Temperature Solid State Synthesis, Characterization, and Application of a Zinc Complex with Pyromellitic Acid

Rong-Gui Yang, Mei-Ling Wang, Ting Liu and Guo-Qing Zhong *

School of Material Science and Engineering, Southwest University of Science and Technology,
Mianyang 621010, China; yangronggui@mails.swust.edu.cn (R.-G.Y.); wangmeiling160206@163.com (M.-L.W.);
liuting0906@mails.swust.edu.cn (T.L.)
* Correspondence: zgq316@163.com or zhongguoqing@swust.edu.cn; Tel.: +86-816-6089371

Received: 31 December 2017; Accepted: 22 January 2018; Published: 24 January 2018

Abstract: The complex [Zn_2(btca)(H_2O)$_4$] was synthesized with 1,2,4,5-benzenetetracarboxylic acid (H_4btca) and zinc acetate as materials via a room-temperature solid state reaction. The composition and structure of the complex were characterized by elemental analyses (EA), Fourier transform infrared spectroscopy (FTIR), X-ray powder diffraction (XRD), and thermogravimetric (TG) analysis. The index results of X-ray powder diffraction data showed that the crystal structure of the complex belonged to monoclinic system with cell parameters a = 9.882 Å, b = 21.311 Å, c = 15.746 Å, and β = 100.69°. In order to expand the application of the complex, the nanometer zinc oxide was prepared by using the complex as a precursor, and the effect of the thermal decomposition temperature on the preparation of the nanometer zinc oxide was studied. The results showed that the grain size of zinc oxide gradually grew with the increase of the pyrolysis temperature, the obtained nanometer zinc oxide was spherical, and the diameter of the particles was about 25 nm.

Keywords: room-temperature solid state reaction; zinc complex; index of X-ray powder diffraction data; precursor; nanometer zinc oxide

---

1. Introduction

Pyromellitic acid, namely 1,2,4,5-benzenetetracarboxylic acid (H_4btca), is usually used as a ligand for the synthesis of the complexes. There are four carboxyl groups in the pyromellitic acid, which leads to a rich variety of coordination patterns and topological structures [1–3]. The tetra-anion of H_4btca is interesting as a suitable ligand for the preparation of inorganic–organic framework structures. At the same time, the existence of aromatic ring in pyromellitic acid is conducive to the transfer of the electrons, which contributes to the preparation of optical, electrical and magnetic chemical materials [4–7]. The common synthetic method for this kind of complexes is the solvothermal method, and the complexes synthesized by this method have protean structures and superior performances [8,9]. However, there are some shortcomings for the method, such as difficult controls of synthetic conditions and difficulty in understanding the reaction process.

The nanophase materials have the distinctive surface effect, volume effect, macroscopic quantum tunneling effect, and so on. As a result, nanomaterials can be applied to the fields of electricity, magnetism, optics, mechanics, catalysis, energy, explosives, and as a combustion improver. The studies and preparation of nano-ZnO have attracted more and more attention globally; for example, the development of nanometer zinc oxide is included in the "863 Plan" in China. There are many methods used to prepare nano-ZnO in laboratory [10–14], which can be divided into chemical methods and physical methods. Among them, chemical methods are more studied, especially the thermal decomposition of zinc complexes with organic acid.

In this paper, the zinc complex of pyromellitic acid was synthesized via a room temperature solid state reaction, and the complex was used as precursor to prepare the nano-ZnO. This synthetic method of this complex is not only easy to operate, but it has a high yield.

2. Experimental Section

2.1. Materials and Physical Measurements

All chemicals purchased were of analytical reagent grade and used without further purification. Zinc acetate ($Zn(Ac)_2 \cdot 2H_2O$) was purchased from Sinopharm Chemical Reagent Co., Ltd. (Shanghai, China) and 1,2,4,5-benzenetetracarboxylic acid (H_4btca) was obtained from Jinan Henghua Sci. and Tec. Co., Ltd. (Jinan, China).

Elemental analyses for C, H, and O in the complex were measured on a Vario EL CUBE elemental analyzer (Elementar, Langenselbold, Germany), and the content of zinc was determined by EDTA complexometric titration with chrome black T as indicator. FTIR spectra were obtained with KBr pellets on a Nicolet 5700 FT-IR spectrophotometer (ARCoptix, Neuchatel, Switzerland) in the range of $4000-400$ cm^{-1}. The powder X-ray diffraction data were collected on a D/max-II X-ray diffractometer (Rigaku, Tokyo, Japan) with Cu $K_{\alpha 1}$ radiation, the voltage of 35 kV, the current of 60 mA, and the scanning speed of $8°$ min^{-1}, in the diffraction angle range of $3-80°$. The thermogravimetric analysis data were obtained using a SDT Q600 thermogravimetry analyzer (TA Instruments, New Castle, DE, USA) in an air atmosphere in the temperature range of $25-800$ °C with a heating rate of 10 °C min^{-1}. Scanning electron microscope (SEM) images of particles were measured on an Ultra 55 field emission scanning electron microscope system (Camcor, Inc., Burlington, NC, USA).

2.2. Synthesis of [Zn₂(btca)(H₂O)₄]

$Zn(Ac)_2 \cdot 2H_2O$ (4.39 g, 20 mmol) and H_4btca (2.54 g, 10 mmol) were weighed and the mixture was placed in a mortar. There was a pungent-smelling gas due to acetic acid generation when grinding the mixture. The mixture was ground for about 30 min at room temperature and then placed in a constant-temperature drying oven with 60 °C. After 10 min, the mixture was continued to be ground. These steps were repeated until there was no pungent-smelling gas produced. The product was a white powder (4.18 g) and the yield was about 92.3%.

2.3. Preparation of Nano-ZnO

The complex [Zn₂(btca)(H₂O)₄] was respectively calcined at 400 °C, 500 °C, and 600 °C for 1.5 h, and the nano-ZnO with a narrow size distribution was obtained. The effect of the calcination temperature on the particle size of nano-ZnO was explored.

3. Results and Discussion

3.1. Composition and Property of the Title Complex

Anal. Calc. (%) for the title complex $C_{10}H_{10}O_{12}Zn_2$: C, 26.51; H, 2.23; O, 42.39; Zn, 28.87. Found (%): C, 26.37; H, 2.27; O, 42.62, Zn, 28.74. The experimental results coincide with the theoretical calculation. The complex consists of two zinc cations, one $btca^{4-}$ anion, and four water molecules. According to the results of FTIR and thermal analysis, the composition of the complex is [Zn₂(btca)(H₂O)₄] (M_r = 452.96). Every Zn(II) ion is coordinated by four oxygen atoms from two carboxyl groups on the same side of the $btca^{4-}$ ligand and two coordinated water molecules, which forms a distorted tetrahedral configuration, and the molecular structure of the complex is shown in Figure 1. The complex is fairly stable at room temperature and has no moisture absorption.

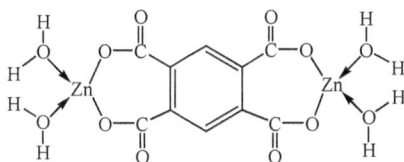

Figure 1. Molecular structure of [Zn$_2$(btca)(H$_2$O)$_4$].

3.2. X-ray Powder Diffraction Analysis

The XRD pattern of the complex is shown in Figure 2. The baseline of the XRD pattern is low and the intensity of the diffraction peak is strong, which indicates that the complex has a good crystalline state. The three main strong peaks appear in $2\theta = 15.44°$, $18.98°$, and $21.88°$ for the complex, and in $2\theta = 25.80°$, $28.97°$, and $11.76°$ for pyromellitic acid (JCPDS No. 13-0882), while in $2\theta = 11.26°$, $19.62°$, and $22.55°$ for zinc acetate (JCPDS No. 21-1467). Comparing with the reactants, the strong peak locations in XRD pattern of the complex are obviously changed. In short, all of these strong peaks in the XRD patterns of the reactants are disappeared in the XRD pattern of the complex. The diffraction angle (2θ), diffractive intensity, and spacing (d) of the resultant are completely different from the reactive materials, which may illuminate that the resultant is a new compound instead of the reactant mixture [15].

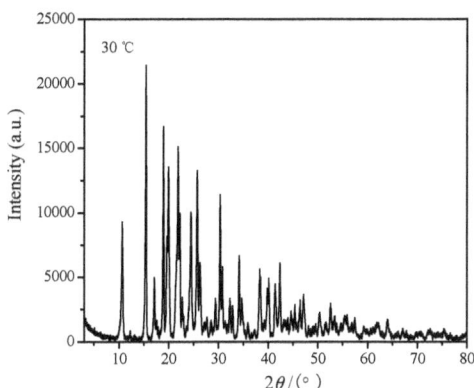

Figure 2. XRD pattern of [Zn$_2$(btca)(H$_2$O)$_4$].

The index calculation of the XRD data bases on the computer program of least squares method [16], and the calculated results are shown in Table 1. All of the XRD data of the complex can be calculated according to the monoclinic system. The calculated interplanar spacing is consistent with the measured value, and the largest relative deviation between the experimental and calculated spacing d_{hkl} is less than 0.3%. As a result, the resultant is a single phase compound, and the crystal structure of the complex belongs to monoclinic system, and its cell parameters are $a = 9.882$ Å, $b = 21.311$ Å, $c = 15.746$ Å, and $\beta = 100.69°$.

Table 1. Experimental data and calculated results for XRD of [Zn$_2$(btca)(H$_2$O)$_4$].

No.	$2\theta/°$	h	k	l	d_{exp}/Å	d_{cal}/Å	I/I_0	No.	$2\theta/°$	h	k	l	d_{exp}/Å	d_{cal}/Å	I/I_0
1	10.64	1	1	−1	8.308	8.300	42.4	29	39.80	1	7	4	2.263	2.263	15.6
2	15.44	1	3	0	5.734	5.733	100.0	30	40.14	2	4	−6	2.245	2.245	20.2
3	17.14	0	0	3	5.169	5.158	21.9	31	41.40	1	7	−5	2.179	2.179	18.2
4	17.61	0	4	1	5.034	5.037	5.9	32	42.36	2	9	−1	2.132	2.132	23.5
5	17.92	1	0	−3	4.946	4.951	3.5	33	43.26	1	8	4	2.090	2.093	3.5
6	18.98	1	4	0	4.672	4.671	77.1	34	43.52	2	2	6	2.078	2.080	3.6
7	19.68	2	0	−2	4.507	4.506	34.4	35	43.96	3	8	0	2.058	2.057	4.2

Table 1. *Cont.*

No.	$2\theta/°$	h	k	l	d_{exp}/Å	d_{cal}/Å	I/I_0	No.	$2\theta/°$	h	k	l	d_{exp}/Å	d_{cal}/Å	I/I_0
8	19.98	2	2	−1	4.440	4.451	61.4	36	44.62	4	6	−1	2.029	2.028	6.7
9	21.54	0	5	1	4.122	4.109	25.1	37	44.90	1	9	−4	2.017	2.017	3.6
10	21.88	1	3	−3	4.059	4.062	68.8	38	45.36	2	6	5	1.998	1.997	9.0
11	22.22	2	3	0	3.997	4.008	44.1	39	46.44	4	5	2	1.954	1.957	9.9
12	22.74	1	5	0	3.907	3.903	13.6	40	47.16	0	1	8	1.926	1.926	14.8
13	22.98	0	0	4	3.867	3.868	6.8	41	48.16	3	6	−6	1.888	1.888	3.2
14	24.44	1	3	3	3.639	3.641	42.4	42	49.58	3	8	3	1.837	1.837	3.5
15	25.72	0	6	−1	3.461	3.462	58.0	43	50.40	3	9	5	1.809	1.809	7.7
16	26.32	1	3	−4	3.383	3.383	24.4	44	51.54	4	8	−3	1.772	1.772	3.2
17	27.24	0	5	3	3.271	3.285	3.7	45	52.60	2	7	6	1.739	1.740	9.9
18	27.74	2	5	−1	3.213	3.216	6.2	46	53.46	0	1	9	1.713	1.714	4.9
19	28.54	1	0	−5	3.125	3.121	4.6	47	55.40	5	7	−1	1.657	1.656	5.2
20	29.44	3	1	1	3.032	3.026	11.2	48	55.88	3	9	4	1.644	1.644	5.3
21	30.38	3	2	1	2.940	2.938	48.5	49	56.84	4	1	6	1.618	1.619	3.1
22	30.82	1	7	0	2.899	2.905	22.2	50	57.46	2	7	7	1.603	1.602	7.1
23	32.34	3	3	−3	2.766	2.767	14.1	51	59.22	1	9	7	1.559	1.560	3.5
24	32.86	1	6	3	2.723	2.723	11.4	52	61.82	2	3	9	1.500	1.499	3.6
25	34.18	0	7	−3	2.621	2.622	29.9	53	62.16	5	7	−6	1.492	1.492	3.4
26	34.70	2	5	3	2.583	2.583	13.5	54	63.96	6	2	3	1.454	1.455	6.3
27	35.92	0	5	5	2.498	2.504	5.3	55	67.04	7	2	−3	1.395	1.395	3.3
28	38.26	1	4	−6	2.350	2.348	25.0	56	72.60	5	4	7	1.301	1.300	3.1

3.3. IR Spectroscopy Analysis

The FTIR spectra of the ligand and the complex [Zn$_2$(btca)(H$_2$O)$_4$] are shown in Figures 3 and 4. Compared with the infrared spectrum of the ligand, the number of infrared absorption bands of the complex is less, and this indicates that the symmetry of the complex is very good. In Figure 4, there is a wide and strong band near 3390 cm^{-1}, which is assigned to the O–H bonds stretching vibration [17]. Compared with the ligand H$_4$btca, the O–H stretching vibration absorption peak of the complex is wider, and the deformation vibration peak of the water molecules appears at 1624 cm^{-1} and there are absorption peaks near 930 cm^{-1} and 760 cm^{-1}, which is an evidence for the existence of coordination water molecules [18]. The stretching vibration peak of the C–H bonds in the benzene ring is detected at 3198 cm^{-1}. The absorption peaks at 1557 cm^{-1} and 1406 cm^{-1} are assigned to the asymmetric and symmetric stretching vibration in the carboxyl groups, respectively [19,20], and the bending vibration peak of the C=O bonds is located at 625 cm^{-1}, all of which are a little lower than the frequency of H$_4$btca (1673 cm^{-1}, 1416 cm^{-1}, and 658 cm^{-1}). Hence, it is illustrated that the carboxyl oxygen atoms in H$_4$btca coordinate with Zn(II) and then cause a redshift [21]. Additionally, the stretching vibration peak of Zn–O bonds detected at 517 cm^{-1} is evidence of the coordination between zinc and oxygen atoms [22].

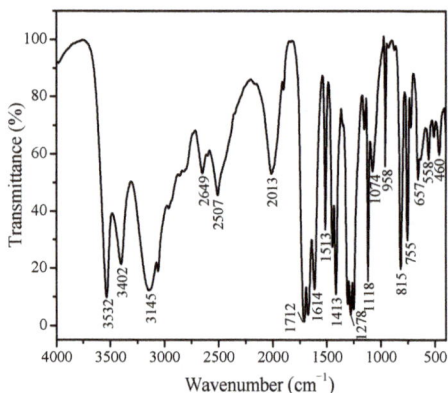

Figure 3. FTIR spectrum of the ligand H$_4$btca.

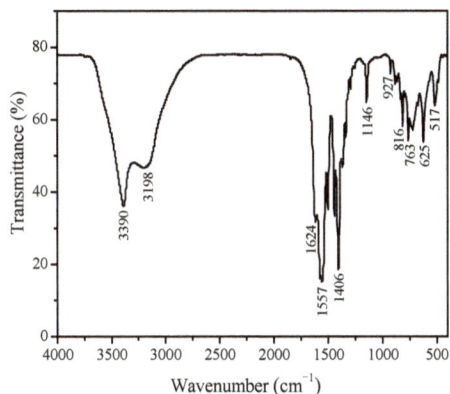

Figure 4. FTIR spectrum of [Zn$_2$(btca)(H$_2$O)$_4$].

3.4. Thermogravimetric Analysis

Studying thermal decomposition of complexes is helpful to understanding of coordination structure and mechanism of thermal decomposition [23]. The thermogravimetric and differential thermogravimetry (TG−DTG) curves of the complex are shown in Figure 5. The TG analysis reveals that the complex is decomposed through two major processes, namely the loss of coordinated water molecules and the oxidative decomposition of the ligand btca^{4-} anion. The first weight loss is approximately 15.77% (calcd. 15.91%) near 110 °C, corresponding to loss of four coordination water molecules. The second weight loss occurred at 488 °C is 48.85%, and the final residue is zinc oxide (Found 35.38%, calcd. 35.93%). The decomposition processes can be expressed by the Equation (1) as follows:

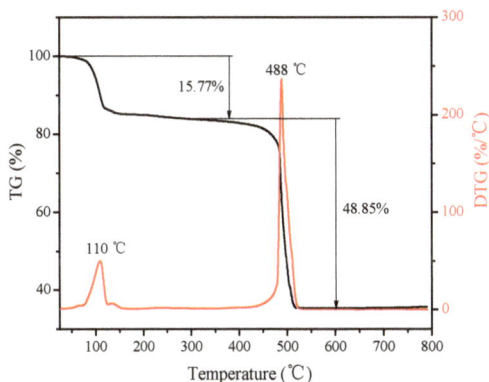

$$[\text{Zn}_2(\text{btca})(\text{H}_2\text{O})_4] \xrightarrow[-4\text{H}_2\text{O}]{110\ °\text{C}} \text{Zn}_2(\text{btca}) \xrightarrow[+6.5\text{O}_2,\ -10\text{CO}_2,\ -\text{H}_2\text{O}]{488\ °\text{C}} 2\text{ZnO} \tag{1}$$

Figure 5. TG−DTG curves of [Zn$_2$(btca)(H$_2$O)$_4$].

3.5. Particle Size and Morphology of Nano-ZnO

In order to expand the application of the title complex, the nanometer zinc oxide was prepared with the complex [Zn$_2$(btca)(H$_2$O)$_4$] as precursor by the thermal decomposition reaction. The complex was respectively calcined at 400 °C, 500 °C, and 600 °C for 1.5 h, and the XRD patterns of the products are shown in Figure 6. At 400 °C, the standard diffraction peaks of the hexagonal system zinc oxide (JCPDS No. 36-1451) is already obtained, while there are some stray peaks, from which we can come to the conclusion that the complex does not decompose completely. This is also consistent with the thermal decomposition temperature of the complex. In the range of nanometer scale, the smaller the grain size is, the wider and weaker the diffraction peaks will be. As can be seen from Figure 6, the diffraction peaks become sharper when the pyrolysis temperature is increased gradually, and this indicates that the grain size of zinc oxide grows gradually and its crystal shape grows more completely. From the calculation by the X-ray single peak Fourier analysis method [24,25], the particle diameters of the nano-ZnO products (a), (b) and (c) are 19 nm, 26 nm, and 32 nm, respectively.

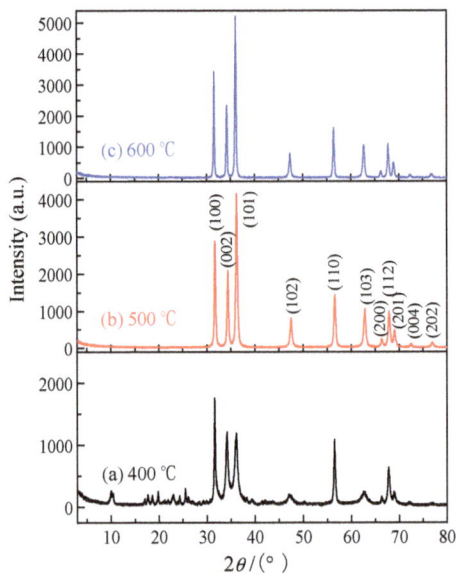

Figure 6. XRD patterns obtained by the calcination of $[Zn_2(btca)(H_2O)_4]$ at different temperatures, (a) 400 °C, (b) 500 °C, and (c) 600 °C.

Figure 7 shows the SEM images of the nano-ZnO obtained from calcining the complex $[Zn_2(btca)(H_2O)_4]$ at 500 °C. It is obvious that the particles are agglomerated together, which is a common phenomenon of nanocrystals. The diameter of the particles is about 25 nm, and it is consistent with the calculated value. The purity of nano-ZnO is high, and the content of zinc oxide determined by EDTA complexometric titration is 99.9%. Therefore, the preparation method of the nano-ZnO has an extensive popularization and application value.

Figure 7. SEM images of nano-ZnO.

4. Conclusions

The complex [Zn_2(btca)(H_2O)$_4$] was synthesized via a solid state reaction at room temperature. The complex was characterized by EA, FTIR, XRD, and TG-DTG. The crystal structure of the complex belonged to monoclinic system with cell parameters a = 9.882 Å, b = 21.311 Å, c = 15.746 Å, and β = 100.69°. Every Zn(II) ion was coordinated by four oxygen atoms from two carboxyl groups on the same side of the btca^{4-} ligand and two coordinated water molecules. The nano-ZnO was prepared with the complex as a precursor by thermal decomposition. This showed that the grain size of zinc oxide gradually grew with the increase of the pyrolysis temperature. The diameter of the nano-ZnO particles prepared at 500 °C is about 25 nm. The particle size of nano-ZnO can be changed by controlling the pyrolysis temperature of the complex precursor.

Acknowledgments: This work was supported by the Longshan academic talent research supporting program of SWUST (No. 17LZX414) and the scientific research funds of Education Department of Sichuan Province (No. 10ZA016).

Author Contributions: Mei-Ling Wang and Rong-Gui Yang tied for the first author. Rong-Gui Yang wrote the manuscript. Mei-Ling Wang synthesized the compound and carried on the analysis to EA, FTIR, and TG-DTG. Ting Liu performed the X-ray powder diffraction analysis. Guo-Qing Zhong conceived and designed the experiments. All authors took part in the writing and discussion processes.

Conflicts of Interest: The authors declare no conflict of interest.

References

1. Xia, C.-K.; Wu, F.; Yang, K.; Wu, Y.-L.; Lu, X.-J. Syntheses, crystal structures and properties of four novel zinc(II) complexes assembled from versatile 1,2,3,5-benzenetetracarboxylic acid and bridging dipyridyl ligands. *Polyhedron* **2016**, *112*, 78–85. [CrossRef]
2. Wang, X.-B.; Lu, W.-G.; Zhong, D.-C. Two zinc(II) metal-organic frameworks with mixed ligands of 5-amino-tetrazolate and 1,2,4,5-benzenetetracarboxylate: Synthesis, structural diversity and photoluminescent properties. *J. Solid State Chem.* **2017**, *250*, 83–89. [CrossRef]
3. Wang, Y.-F.; Li, Z.; Sun, Y.-C.; Zhao, J.-S.; Zhang, S.-C. Synthesis, characterization and crystal structures of 2D Co.(II)/Zn(II)-coordination polymers containing 3-(pyridin-4-yl)-5-(pyrazin-2-yl)-1H-1,2,4-triazole and benzenetetracarboxylate co-ligands. *Inorg. Chem. Commun.* **2014**, *44*, 25–28. [CrossRef]
4. Sanda, S.; Goswami, S.; Jena, H.S.; Parshamoni, S.; Konar, S. A family of three magnetic metal organic frameworks: Their synthesis, structural, magnetic and vapour adsorption study. *CrystEngComm* **2014**, *16*, 4742–4752. [CrossRef]
5. Zhao, Y.; Lu, H.; Yang, L.; Luo, G.G. Reversible adsorption of a planar cyclic (H_2O)$_6$ cluster held in a 2D Cu^{II}-coordination framework. *J. Mol. Struct.* **2016**, *1088*, 155–160. [CrossRef]
6. Wang, X.-X.; Zhang, M.-X.; Yu, B.; Hecke, K.V.; Cui, G.-H. Synthesis, crystal structures, luminescence and catalytic properties of two d^{10} metal coordination polymers constructed from mixed ligands. *Spectrochim. Acta A* **2015**, *139*, 442–448. [CrossRef] [PubMed]
7. Xia, C.; Yang, K.; Wu, F.; Lu, X.; Min, Y.; Xie, J. Syntheses and characterization of two new compounds based on versatile 1,2,3,5-benzenetetracarboxylic acid and 2,2'-bibenzimidazole. *J. Coord. Chem.* **2017**, *70*, 2510–2519. [CrossRef]
8. Du, S.; Ji, C.; Xin, X.; Zhuang, M.; Yu, X.; Lu, J.; Lu, Y.; Sun, D. Syntheses, structures and characteristics of four alkaline-earth metal-organic frameworks (MOFs) based on benzene-1,2,4,5-tetracarboxylicacid and its derivative ligand. *J. Mol. Struct.* **2017**, *1130*, 565–572. [CrossRef]
9. Zhong, G.-Q.; Li, D.; Zhang, Z.-P. Hydrothermal synthesis, crystal structure and magnetic property of a homodinuclear ternary coordination polymer of nickel(II). *Polyhedron* **2016**, *111*, 11–15. [CrossRef]
10. Wang, G.; Ma, D.; Jia, X.; Cui, X.; Zhao, X.; Wang, Y. In situ preparation of nanometer-scale zinc oxide from zinc acetate in the reaction for the synthesis of dimethyl toluene dicarbamate and its catalytic decomposition performance. *Ind. Eng. Chem. Res.* **2016**, *55*, 8011–8017. [CrossRef]
11. Lee, T.-H.; Ryu, H.; Lee, W.-J. Fast vertical growth of ZnO nanorods using a modified chemical bath deposition. *J. Alloys Compd.* **2014**, *597*, 85–90. [CrossRef]

12. Kamari, H.M.; Al-Hada, N.M.; Saion, E.; Shaari, A.H.; Talib, Z.A.; Flaifel, M.H.; Ahmed, A.A.A. Calcined solution-based PVP influence on ZnO semiconductor nanoparticle properties. *Crystals* **2017**, *7*, 2. [CrossRef]

13. Hussain, S.; Liu, T.; Kashif, M.; Cao, S.; Zeng, W.; Xu, S.; Naseer, K.; Hashim, U. A simple preparation of ZnO nanocones and exposure to formaldehyde. *Mater. Lett.* **2014**, *128*, 35–38. [CrossRef]

14. Thilagavathi, T.; Geetha, D. Nano ZnO structures synthesized in presence of anionic and cationic surfactant under hydrothermal process. *Appl. Nanosci.* **2014**, *4*, 127–132. [CrossRef]

15. Zhong, G.-Q.; Zhong, Q. Solid-solid synthesis, characterization, thermal decomposition and antibacterial activities of zinc(II) and nickel(II) complexes of glycine-vanillin Schiff base ligand. *Green Chem. Lett. Rev.* **2014**, *7*, 236–242. [CrossRef]

16. Li, D.; Zhong, G.-Q.; Wu, Z.-X. Solid-solid synthesis, characterization and thermal decomposition of a homodinuclear cobalt(II) complex. *J. Serb. Chem. Soc.* **2015**, *80*, 1391–1397. [CrossRef]

17. Tai, X.-S.; Meng, Q.-G.; Liu, L.-L. Synthesis, crystal structure, and cytotoxic activity of a novel eight-coordinated dinuclear Ca(II)-Schiff base complex. *Crystals* **2016**, *6*, 109. [CrossRef]

18. Nakamoto, K. *Infrared and Raman Spectra of Inorganic and Coordination Compounds*, 6th ed.; John Wiley & Sons Inc.: Hoboken, NJ, USA, 2009.

19. Majumder, A.; Gramlich, V.; Rosair, G.M.; Batten, S.R.; Masuda, J.D.; El Fallah, M.S.; Ribas, J.; Sutter, J.P.; Desplanches, C.; Mitra, S. Five new cobalt(II) and copper(II)-1,2,4,5-benzenetetracarboxylate supramolecular architectures: Syntheses, structures, and magnetic properties. *Cryst. Growth Des.* **2006**, *6*, 2355–2368. [CrossRef]

20. Lei, R.; Zhang, H.-H.; Hu, J.; Chai, X.-C.; Zhang, S.; Li, C.-X.; Chen, Y.-P.; Sun, Y.-Q. Synthesis and single-crystal structure of a silver(I) carboxyarylphosphonate: [Ag(H$_2$BCP)(4,4′-bipy)]·2H$_2$O. *Chin. J. Struct. Chem.* **2010**, *29*, 655–659.

21. Tai, X.S.; You, H.Y. A new 1D chained coordination polymer: Synthesis, crystal structure, antitumor activity and luminescent property. *Crystals* **2015**, *5*, 608–616. [CrossRef]

22. Bahnasawy, R.M.E.; El-Tabl, A.S.; Shakdofa, M.M.E.; El-Wahed, N.M.A. Cu(II), Ni(II), Co.(II), Mn(II), Zn(II) and Cd(II) complexes of ethyl-3-(2-carbamothioylhydrazono)-2-(hydroxyimino)butanoate: Synthesis, characterization and cytotoxicity activity. *Chin. J. Inorg. Chem.* **2014**, *30*, 1435–1450.

23. Zhong, G.Q.; Shen, J.; Jiang, Q.Y.; Jia, Y.Q.; Chen, M.J.; Zhang, Z.P. Synthesis, characterization and thermal decomposition of Sb^{III}–M–Sb^{III} type trinuclear complexes of ethylenediamine-N,N,N′,N′-tetraacetate (M:Co(II), La(III), Nd(III), Dy(III)). *J. Therm. Anal. Calorim.* **2008**, *92*, 607–616. [CrossRef]

24. Deacon, G.B.; Phillips, R.J. Relationships between the carbon-oxygen stretching frequencies of carboxylato complexes and the type of carboxylate coordination. *Coord. Chem. Rev.* **1980**, *33*, 227–250. [CrossRef]

25. Langford, J.I. A rapid method for analysing the breadths of diffraction and spectral lines using the Voigt function. *J. Appl. Cryst.* **1978**, *11*, 10–14. [CrossRef]

crystals

MDPI

Article

Morphology Transition of ZnO Nanorod Arrays Synthesized by a Two-Step Aqueous Solution Method

Guannan He [1,2,*], Bo Huang [3], Zhenxuan Lin [1,4,*], Weifeng Yang [5], Qinyu He [1,2] and Lunxiong Li [1]

[1] Guangdong Provincial Key Laboratory of Quantum Engineering and Quantum Materials, South China Normal University, Guangzhou 510006, Guangdong, China; gracylady@163.com (Q.H.); lilx@m.scnu.edu.cn (L.L.)

[2] Guangdong Engineering Technology Research Center of Efficient Green Energy and Environmental Protection Materials, Guangzhou 510006, Guangdong, China

[3] Department of Electronic Engineering, Jinan University, Guangzhou 510632, Guangdong, China; abohuang@gmail.com

[4] Whiting School of Engineering, Johns Hopkins University, 3400 North Charles Street, Baltimore, MD 21218-2608, USA

[5] Institute of Materials Research and Engineering, A*STAR (Agency for Science, Technology and Research), 2 Fusionopolis Way, Innovis, Singapore 138634, Singapore; Yangw@imre.a-star.edu.sg

* Correspondence: hegn@m.scnu.edu.cn (G.H.); zlin22@jhu.edu (Z.L.); Tel.: +86-20-3931-0066 (G.H.); +1-443-858-2933 (Z.L.); Fax: +86-20-3931-0882 (G.H.)

Received: 21 February 2018; Accepted: 27 March 2018; Published: 30 March 2018

Abstract: ZnO nanorod arrays (ZNAs) with vertically-aligned orientation were obtained by a two-step aqueous solution method. The morphology of the ZnO nanorods was regulated by changing the precursor concentration and the growth time of each step. ZnO nanorods with distinct structures, including flat top, cone top, syringe shape, and nail shape, were obtained. Moreover, based on the X-ray diffraction (XRD) and the transmission electron microscope (TEM) analysis, the possible growth mechanisms of different ZnO nanostructrues were proposed. The room-temperature PL spectra show that the syringe-shaped ZNAs with ultra-sharp tips have high crystalline quality. Our study provides a simple and repeatable method to regulate the morphology of the ZNAs.

Keywords: ZnO nanorod arrays; aqueous solution method; growth mechanism; PL spectra

1. Introduction

In recent years, one-dimensional nanostructures, like nanorods, nanotubes, and nanowires, have attracted much more attention, due to their potential applications on the nanodevices, such as sensors, field emission, and detectors, etc. [1–8]. Among them, ZnO, as an important II–VI semiconductor with a direct band gap of 3.37 eV, larger exciton binding energy of 60 mV [9,10] at room temperature, and piezoelectricity [11], has attracted special interest. Especially, vertically aligned ZnO nanorod arrays (ZNAs), due to their non-toxicity, low-cost, large surface to volume ratio, and ease of large scale fabrication, can be employed in many devices, like solar cells, UV lasers, gas sensors, UV detectors, and photocatalysis [12–20]. Various methods have been used to synthesize ZNAs, including chemical vapor deposition (CVD) [21,22], pulsed laser deposition (PLD) [23], molecular beam epitaxy (MBE) [10,24], etc. Compared to these methods, the aqueous solution method [25,26] has emerged more recently, which paves a facile way to obtain preferred-aligned ZNAs on various substrates just by pre-coating a ZnO seed layer. Furthermore, the aqueous solution method is always conducted under low temperature (<95 °C), which extends its application on the temperature-sensitive substrates. However, until now, exact control over the morphologies of ZnO nanostructure during the solution growth is still a challenge, especially for some sophisticated structures. Furthermore, the growth mechanisms of ZnO nanostructures in solution are far from fully understood.

In this article, we proposed a two-step aqueous solution method to obtain preferred orientation ZNAs. The morphologies of the ZnO nanorods were regulated by changing the precursor concentration and the growth time in each step. ZnO nanorods with distinct structures, including flat top, cone top, syringe shape, and nail shape, were obtained. Moreover, based on the X-ray diffraction (XRD) and the transmission electron microscope (TEM) characterizations, the possible growth mechanisms of different ZnO nanostructrues were proposed. Our study allows to understand the growth mechanisms of the ZNAs with different morphologies in depth, and provides a simple and repeatable method to regulate the morphology of the ZNAs.

2. Materials and Methods

2.1. Substrate Preparation

Multi-crystalline silicon (mc-Si) wafers (~200 μm thickness) were selected as the substrates. The wafers were first etched to remove the saw damage in alkaline solution. Subsequently, a layer of SiN_x was deposited on the front surface of the wafer by plasma enhanced chemical vapor deposition (PECVD), and the thickness was about 80 nm. Then the substrates were cut into ~50 mm × 50 mm pieces by laser and cleaned by acetone, ethanol, and deionized (DI) water successively, and dried by nitrogen gas.

2.2. Seed Layer Deposition

The ZnO seed layer was deposited on the SiN_x surface using the sol-gel method. Firstly, zinc acetate dehydrate and equivalent molar monoethanolamine were dissolved in ethyl alcohol, then stirred for 2 h and kept at room temperature for 24 h to obtain a homogenous and clear sol (0.005 M). A few drops of sol were dripped onto the substrates and kept for a few seconds. Then the substrates were rinsed with ethyl alcohol and kept in a vacuum oven at 108 °C for 10 min. The former steps were repeated five times and then the gel-coated substrates were annealed in a furnace at 300 °C for 1 h to obtain a uniform ZnO seed layer on the substrate.

2.3. ZNAs Fabrication

The ZNAs were synthesized by a two-step aqueous solution method. The solution contained equal concentrations (0.01 M or 0.05 M) of zinc nitrate hexahydrate ($Zn(NO_3)_2 \cdot 6H_2O$) and methenamine ($C_6H_{12}N_4$) as the precursors, and DI water as the solvent. Each solution was mixed thoroughly by magnetic stirring for 2 h, and then placed at room temperature for 24 h to obtain a homogeneous and clear solution. The substrates prepared above were divided into two groups for the following experiments. In the first step (T1), the substrates of Group 1 were kept face down and immersed into the solution with precursor concentration of 0.05 M at 95 °C for 2 h. After this treatment, the substrates were withdrawn from the beaker and rinsed thoroughly with DI water. The sample was labeled as H-0. In the second step (T2), five samples were chosen from the samples after T1 treatment, kept face down and immersed into the second solution with precursor concentrations of 0.01 M at 95 °C for 1 h, 2 h, 4 h, and 10 h, respectively. Other growth conditions remained unchanged. After all these treatments, the samples were rinsed with DI water thoroughly and dried at room temperature. The samples were labeled as H-1, H-2, H-4, and H-10, respectively. Meanwhile, two substrates of Group 2 were immersed in the solution with precursor concentration of 0.01 M in T1 for 2 h. Then they were withdrawn from the beaker and rinsed thoroughly with DI water. One of them was labeled as L-0, and the other was immersed into the solution with precursor concentration of 0.05 M in T2 for another 2 h, and other conditions remained the same. This sample was labeled as L-2. The growth conditions of the samples are shown in Table 1.

Table 1. Growth conditions of the ZNAs.

Sample	Growth Temperature (°C)	T1		T2	
		Precursor Concentration (M)	Growth Time (h)	Precursor Concentration (M)	Growth Time (h)
H-0				—	—
H-1		0.05			1
H-2	95		2	0.01	2
H-4					4
H-10					10
L-0		0.01		—	—
L-2				0.05	2

2.4. Characterization

The morphologies of the ZNAs were characterized by scanning electron microscopy (SEM) (HITACHI SU8010, Tokyo, Japan) and transmission electron microscope (TEM) (JEM-2100HR, Tokyo, Japan). The crystal structure and crystallographic properties of the samples were analyzed using X-ray diffraction (XRD) (RIGAKU D/MAX2200, Tokyo, Japan) with Cu-Kα as the radiation source within the 2θ range of 20–60° at a scan rate of 0.0167° step^{-1}. The room-temperature photoluminescence (PL) properties were characterized using fluorescence spectrophotometer (HITACHI F-4500, Tokyo, Japan) with an excitation wavelength of 345 nm.

3. Results and Discussion

Figure 1 shows the morphologies of the samples of Group 1 with precursor concentration of 0.05 M in T1, and 0.01 M in T2 with different growth times. Figure 1a is the SEM image of the sample H-0, which was synthesized at 95 °C, with a precursor concentration of 0.05 M for 2 h in T1 and without T2 treatment. The nanorods are homogeneously distributed and perpendicular to the substrate. The diameter seems to be uniform along the nanorod and, as shown in the top view image (inset of Figure 1a), almost every nanorod is terminated with a flat hexagonal top and the average diameter is about 50 nm. Figure 1b,c are the cross-section images of the samples H-1 and H-2, respectively. As the time of T2 treatment increases, the diameter of the nanorod tapered on the top, like a cone shape. When the time of T2 increased to 4 h for the sample H-4, almost every nanorod is terminated with an ultra-sharp tip like a syringe, as shown in Figure 1d. The diameter of the needle tips is much smaller than that of the original nanorods. As the time of T2 continuously increased to 10 h, the lengths of the needle tips grow up and eventually transform to nanowires, as shown in the cross-section (Figure 1e) and the top view (Figure 1f) images of sample H-10. The nanowires originate from the tops of the prepared nanorods in T1 and are twinned with each other on the ends. The total length is about several microns.

Regarding the samples of Group 2, Figure 2a,b are the top view and cross-section SEM images of sample L-0, which was synthesized with 0.01 M precursor concentration in T1 for 2 h, and without T2 treatment. The ZnO nanorods are perpendicular to the substrate, and the diameter is almost uniform along the nanorod. However, the average diameter is around 32 nm, much thinner than sample H-0, as shown in Figure 1a, which was synthesized in a high precursor concentration. Figure 2c,d are the top view and cross-section SEM images of sample L-2, which was synthesized with 0.01 M precursor concentration in T1 for 2 h and 0.05 M precursor concentration in T2 for 2 h. As shown in the top view image, the nanorods exhibit a regular hexagon on the tops. The average diameter of the tops is about 58 nm. From the cross-section image, the nanorods are also perpendicular to the substrate, but the diameter is not uniform along each single rod. Instead, the diameter is very thin on the bottom, and increases from the bottom to the top along the nanorod, reaching its highest value on the top, like a nail shape.

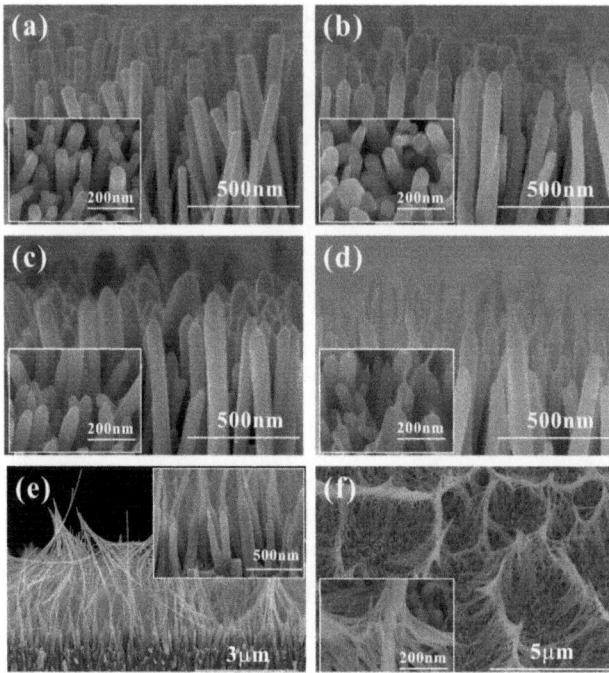

Figure 1. The cross-section and top view (inset) images of the ZNAs synthesized with 0.05 M precursor concentration in T1 for 2 h and 0.01 M in T2 for: (**a**) 0 h (H-0); (**b**) 1 h (H-1); (**c**) 2 h (H-2); (**d**) 4 h (H-4); and (**e**) the cross-section and (**f**) the top view images of the ZNAs synthesized with 0.05 M precursor concentration in T1 for 2 h and 0.01 M in T2 for 10 h (H-10). The insets are their high-magnification images.

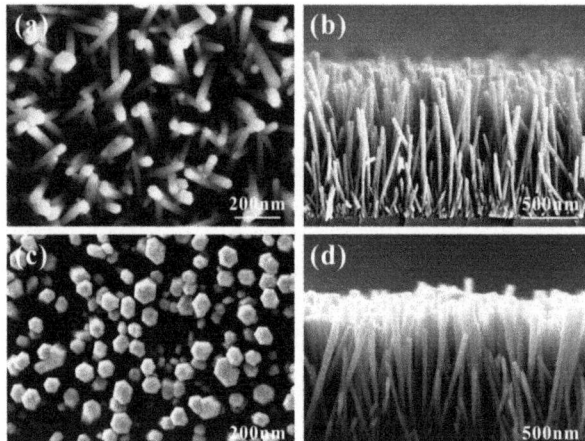

Figure 2. (**a,b**) The top-view and cross-section SEM images of the sample L-0 which was synthesized with 0.01 M precursor concentration in T1 for 2 h without T2 treatment. (**c,d**) The top view and cross-section SEM images of the sample L-2 which was synthesized with 0.01 M precursor concentration in T1 for 2 h, and with 0.05 M precursor concentration in T2 for 2 h.

Figure 3 shows the XRD patterns of the sample H-0, H-4, L-0, L-2, and the seed layer. The diffraction peaks correspond to the ZnO hexagonal wurtzite structure (JCPDS 36-1451). For each sample, only one diffraction peak was found at 34.54°, which indicates a prefered (002) orientation. No other diffraction peaks were detected, which indicates the high purity of the samples.

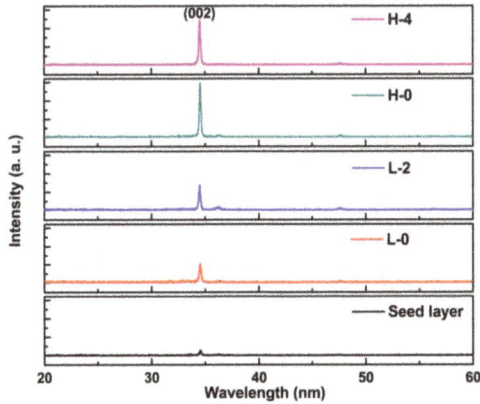

Figure 3. The XRD patterns of the seed layer and the sample L-0, L-2, H-0, and H-4.

In order to further understand the growth mechanisms of the ZnO nanostructures, the TEM analysis was conducted on sample H-4. As shown in the low-resolution TEM images of Figure 4a,b, the syringe shape with obvious change in diameter is clearly exhibited in the images. Figure 4c,d are the high-resolution TEM (HRTEM) images of the corresponding selected areas of the red boxes in Figure 4b, which are the ultra-sharp tip and the thick bottom rod, respectively. The lattice spacing of 0.26 nm corresponds to the ZnO (0001) plane. Inset of Figure 4d is the corresponding selected-area electron diffraction (SAED) pattern. The results of the HRTEM and the SAED pattern indicate that the syringe-shaped ZnO nanorod grows along the [0001] direction, which corresponds to the XRD results.

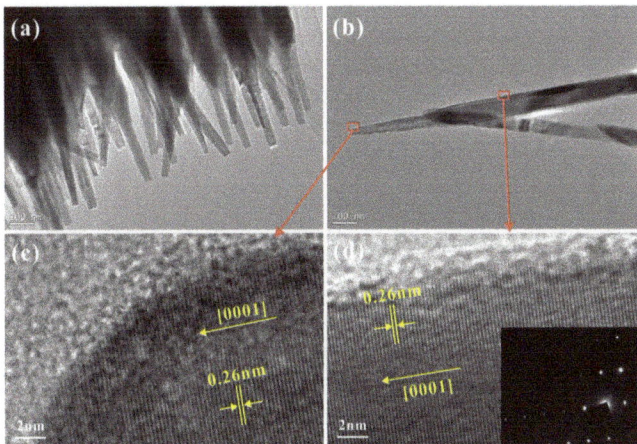

Figure 4. (**a,b**) The TEM images of the syringe-shaped ZNAs of the sample H-4. (**c,d**) HRTEM images of the selected areas marked on the syringe-shaped ZnO nanorod shown in (**b**). Inset of (**d**) is the corresponding SAED pattern.

In the aqueous solution condition, the ZnO crystals were obtained according to the following reactions. Initially, when $Zn(NO_3)_2 \cdot 6H_2O$ was dissolved into the water, zinc ion (Zn^{2+}) and nitrate ion (NO_3^-) were dissociated as the following Equation (1):

$$Zn(NO_3)_2 \cdot 6H_2O + H_2O \rightarrow Zn^{2+} + 2NO_3^- + 7H_2O \tag{1}$$

Methenamine is believed to act as a stabilizer, which can slowly hydrolyze in water and gradually produce ammonium ions (NH_4^+) and hydroxide ions (OH^-) [27,28]:

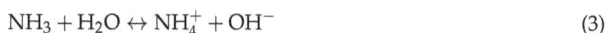

$$(CH_2)_6N_4 + 6H_2O \rightarrow 6HCHO + 4NH_3 \tag{2}$$

$$NH_3 + H_2O \leftrightarrow NH_4^+ + OH^- \tag{3}$$

$Zn(OH)_4{}^{2-}$ and $Zn(NH_3)_4{}^{2+}$ ions are generally assumed as growth units which are formed after mixing and heating the precursor solutions as Equations (4)–(6) [29–31]. The ZnO crystal can be formed by dehydration of $Zn(OH)_4{}^{2-}$ and $Zn(NH_3)_4{}^{2+}$ as Equations (7) and (8):

$$Zn^{2+} + 2OH^- \rightarrow Zn(OH)_2 \tag{4}$$

$$Zn(OH)_2 + 2OH^- \rightarrow Zn(OH)_4^{2-} \tag{5}$$

$$Zn^{2+} + 4NH_4^+ \rightarrow Zn(NH_3)_4^{2+} \tag{6}$$

$$Zn(OH)_4^{2-} \rightarrow ZnO + H_2O + 2OH^- \tag{7}$$

$$Zn(NH_3)_4^{2+} + 2OH^- \rightarrow ZnO + 4NH_3 + H_2O \tag{8}$$

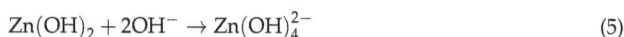

The precursor concentration directly influences the amount of Zn^{2+} and OH^-. At the beginning of the reaction, the pH values of 0.05 M and 0.01 M solution are 6.6 and 6.8, respectively. When the reaction continues for 1 h, 2 h, and 4 h, the pH value of 0.05 M solution remains constant at 5.3. The large decrease in the pH value may due to the reaction of Zn^{2+} and OH^-, resulting in a large consumption of OH^-. As for the 0.01 M solution, when the reaction continues for 1 h, 2 h, 4 h, and 10 h, the pH values are 5.6, 5.7, 6.1, and 6.3. At the early stage of the reaction, the pH value decreases, which is similar to the 0.05 M solution. However, as the reaction proceeds, the pH value increases slowly. This is because the gradually-released OH^- ions are not consumed owing to the low amount of Zn^{2+} ions remaining in the solution after a period of growth.

The ZnO seed layer plays a vital role on the alignment of the ZnO nanorods. As shown in the XRD pattern (Figure 3, black line), the seed layer exhibits only a (002) diffraction peak, which indicates its high orientation with c-axis perpendicular to the substrate. Due to the natural properties of the polar crystal, the polar surfaces can be easily charged positively or negatively [25]. Thus, the ZnO seed layer surface consists of a positively-charged polar Zn^{2+}-terminated (001) plane or a negatively-charged polar O^{2-}-terminated $(00\bar{1})$ plane [30,32]. Then the surface can attract growth units of $Zn(NH_3)_4{}^{2+}$ or $Zn(OH)_4{}^{2-}$, which can be dehydrated to form ZnO crystal. Continuously, the polar surfaces absorb the growth units and repeat dehydration to form ZnO crystal until the supply of growth units stops. Thus, the crystal growth rate is faster along the c-axis than other nonpolar surfaces [25]. It is easy to obtain ZnO nanorod arrays with high alignment perpendicular to the substrate with the ZnO seed layer.

The growth mechanisms of the syringe-shaped and the nail-shaped ZNAs are illustrated in Figure 5. The precursor concentration can influence the relative growth rate of different orientations, which is the key factor in determining the overall morphology of the ZnO crystal. Moreover, the growth rate of non-polar facets {0110} is more sensitive to the precursor concentration change than polar facets {0001}, which has also been detected by Zhang et al. [33]. In the T1 step, the supply of the growth units is stable during the early stage of crystal growth and, accordingly, the growth rate ratio of {0001} facets and {0110} facets changes very slightly. Thus, ZnO nanorods terminated with flat hexagonal tops were obtained, as shown in the sample H-0 (Figure 1a). If the precursor concentration in T2 is lower than that in T1, as shown in Figure 5a, the growth rate of {0110} slows down, but pieces by laser and clean crystal maintains a relatively high growth rate on {0001} facets. The tops of the original nanorods serve as energetically-favored sites, so ZnO nanorods with cone-shaped tops were obtained, as shown in the sample H-1 and H-2 (Figure 1b,c). When the growth rate of {0110} decreases to a critical value, the growth rate ratio of {0110} and {0001} becomes stable in the current solution, and accordingly the nanorods will grow with uniform diameter again. Thus, syringe-shaped nanorods were obtained when the time of T2 extended to 4 h, as shown in the sample H-4 (Figure 1d). If the precursor concentration in T2 is higher than that in T1, as shown in Figure 5b, the growth rate ratio of {0110} and {0001} increases. However, confined by the nucleation sites, the diameter cannot increase suddenly, so the diameter of the nanorods enhances layer by layer until the growth rate ratio of {0110} and {0001} return to stable, nail-shaped ZNAs with inverted hexagonal pyramid tips forming, as shown in the sample L-2 (Figure 2c,d).

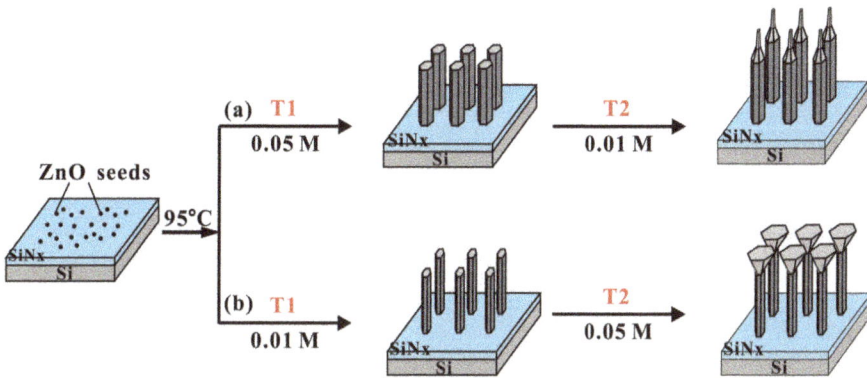

Figure 5. Schematic of the growth mechanisms of: (**a**) the syringe-shaped ZNAs; and (**b**) the nail-shaped ZNAs.

Figure 6 shows the room-temperature PL spectra of the sample H-0, H-4, L-0, and L-2. All the samples exhibit a UV emission peak located at around 384 nm, which is attributed to the recombination of the free excitons (electrons from the conduction band and holes from the valence band) [34–36]. The sample H-4 shows a strong and sharp UV emission indicating its high crystalline. The green emissions are found ranging from 500 nm to 600 nm which are mainly attributed to the defects, such as O-vacancy, Zn-interstitial, and absorbed molecules [30,31,37]. O-vacancy has three different charged states: neutralized oxygen vacancy (V_O^*), single-ionized oxygen vacancy (V_O^+), and doubly-ionized vacancy (V_O^{++}). The electrons from the conduction band (CB) recombine with the doubly-ionized vacancies (V_O^{++}), which generate green emission. The single-ionized oxygen vacancies (V_O^+) capture electrons from CB, and then recombine with the photo-generated holes in the valance band (VB), which always generates a green emission. It is commonly believed that the oxygen-related species, such as O_2 and OH^- can be absorbed on the surface of ZnO nanorods. The electrons or holes captured by the O_2 and OH^- on the surface of ZnO nanorods can nonradiatively recombine [37]. The large surface-to-volume ratio will induce a high amount of O_2 and OH^- absorbed on the surface, which can capture more photo-generated electrons or holes. Nonradiative recombinations increase and, as a result, induce the decrease of green emissions. For our samples, H-4 with ultrasharp tips has the highest surface-to-volume ratio and, accordingly has the lowest green emission. The sample H-0, with the lowest surface-to-volume ratio, exhibits the highest green emission. L-0 and L-2 are in the middle levels.

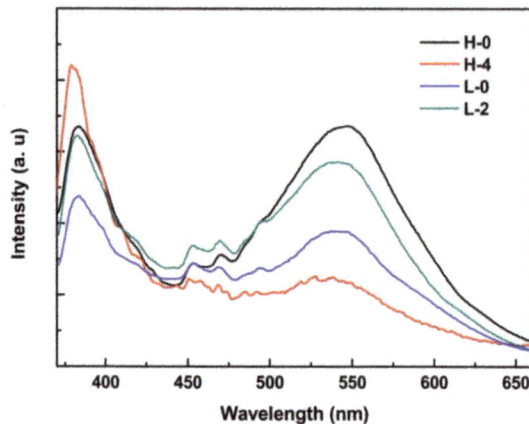

Figure 6. Room-temperature PL spectra of the sample H-0, H-4, L-0, and L-2.

4. Conclusions

In conclusion, the precursor concentration and the growth time in each step are the key factors to determine the morphology of the ZnO nanorods. By changing the growth conditions, ZnO nanorods with different morphologies, including flat top, cone top, syringe-shaped, and nail-shaped, were obtained. The growth mechanism of each nanostructure was discussed. The room-temperature PL spectra show that the syringe-shaped ZNAs with ultra-sharp tips have higher intensity of UV emission peak, indicating their high crystalline quality.

Acknowledgments: The work was financially supported by The Provincial Natural Science Foundation of Guangdong Province, China (no. 2017A030310064), and The National Natural Science Foundation of China (no. 61307080, no. 61404052, no. 51372092, no. 51672090).

Author Contributions: Guannan He characterized the samples, analyzed the data and wrote the article. Zhenxuan Lin performed the experiments and obtained ZnO nanorod arrays with different structures. Bo Huang and Weifeng Yang contributed in the discussion of the experimental results and proposed growth models. Qinyu He and Lunxiong Li contributed in the revision of the paper.

Conflicts of Interest: The authors declare no conflict of interest.

References

1. Choi, S.; Bonyani, B.; Sun, G.J.; Lee, J.K.; Hyun, S.K.; Lee, C. Cr_2O_3 nanoparticle-functionalized WO_3 nanorods for ethanol gas sensors. *Appl. Surf. Sci.* **2018**, *432*, 241–249. [CrossRef]
2. Long, H.W.; Li, Y.Q.; Zeng, W. Substrate-free synthesis of WO_3 nanorod arrays and their superb NH_3-sensing performance. *Mater. Lett.* **2017**, *209*, 342–344. [CrossRef]
3. Yin, Z.Z.; Cheng, S.W.; Xu, L.B.; Liu, H.Y.; Huang, K.; Li, L.; Zhai, Y.Y.; Zeng, Y.B.; Liu, H.Q.; Shao, Y.; et al. Highly sensitive and selective sensor for sunset yellow based on molecularly imprinted polydopmine-coated multi-walled carbon nanotubes. *Biosens. Bioelectron.* **2018**, *100*, 565–570. [CrossRef] [PubMed]
4. Wang, Z.H.; Yang, C.C.; Yu, H.C.; Yeh, H.T.; Peng, Y.M.; Su, Y.K. Electron field emission enhancement based on Al-doped ZnO nanorod arrays with UV exposure. *IEEE Trans. Electron Devices* **2018**, *65*, 251–256. [CrossRef]
5. Hou, J.W.; Wang, B.B.; Ding, Z.J.; Dai, R.C.; Wang, Z.P.; Zhang, Z.M.; Zhang, J.W. Facile fabrication of infrared photodetector using metastable vanadium dioxide VO_2 (B) nanorod networks. *Appl. Phys. Lett.* **2017**, *111*, 072107. [CrossRef]
6. Lee, S.; Lee, W.Y.; Jang, B.; Kim, T.; Bae, J.H.; Cho, K.; Kim, S.; Jang, J. Sol-gel processed p-type CuO phototransistor for a near-infrared sensor. *IEEE Electron Device Lett.* **2018**, *39*, 47–50. [CrossRef]

7.   Li, Y.M.; Zhang, Q.S.; Niu, L.Y.; Liu, J.; Zhou, X.F. TiO$_2$ nanorod arrays modified with SnO$_2$-Sb$_2$O$_3$ nanoparticles and application in perovskite solar cell. *Thin Solid Films* **2017**, *621*, 6–11. [CrossRef]
8.   Momeni, M.M.; Ghayeb, Y.; Menati, M. Facile and green synthesis of CuO nanoneedles with high photo catalytic activity. *J. Mater. Sci. Mater. Electron.* **2016**, *27*, 9454–9460. [CrossRef]
9.   Özgür, Ü.; Alivov, Y.I.; Liu, C.; Teke, A.; Reshchikov, M.A.; Doğan, S.; Avrutin, V.; Cho, S.-J.; Morkoç, H. A comprehensive review of ZnO materials and devices. *J. Appl. Phys.* **2005**, *98*, 041301. [CrossRef]
10. Montenegro, D.N.; Souissi, A.; Martinez-Tomas, C.; Munoz-Sanjose, V.; Sallet, V. Morphology transtions in ZnO nanorods grown by MOCVD. *J. Cryst. Growth* **2012**, *359*, 122–128. [CrossRef]
11. Lu, S.N.; Qi, J.J.; Wang, Z.Z.; Lin, P.; Liu, S.; Zhang, Y. Size effect in a cantilevered ZnO micro/nanowire and its potential as a performance tunable force sensor. *RSC Adv.* **2013**, *3*, 19375–19379. [CrossRef]
12. Yuan, Z.L.; Yao, J.C. Growth of well-aligned ZnO nanorod arrays and their application for photovoltaic devices. *J. Electron. Mater.* **2017**, *46*, 6461–6465. [CrossRef]
13. Lung, C.M.; Wang, W.C.; Chen, C.H.; Chen, L.Y.; Chen, M.J. ZnO/Al$_2$O$_3$ core/shell nanorods array as excellent anti-reflection layers on silicon solar cells. *Mater. Chem. Phys.* **2016**, *180*, 195–202. [CrossRef]
14. Fujiwara, H.; Suzuki, T.; Niyuki, R.; Sasaki, K. ZnO nanorod array random lasers fabricated by a laser-induced hydrothermal synthesis. *New J. Phys.* **2016**, *18*, 103046. [CrossRef]
15. Harale, N.S.; Kamble, A.S.; Tarwal, N.L.; Mulla, I.S.; Rao, V.K.; Kim, J.H.; Patil, P.S. Hydrothermally grown ZnO nanorods arrays for selective NO$_2$ gas sensing: Effect of anion generating agents. *Ceram. Int.* **2016**, *42*, 12807–12814. [CrossRef]
16. Kim, D.; Kim, W.; Jeon, S.; Yong, K. Highly efficient UV-sensing properties of Sb-doped ZnO nanorod arrays synthesized by a facile, singlestep hydrothermal reaction. *RSC Adv.* **2017**, *7*, 40539–40548. [CrossRef]
17. Husham, M.; Hamidon, N.M.; Paiman, S.; Abuelsamen, A.A.; Farhat, O.F.; Al-Dulaimi, A.A. Synthesis of ZnO nanorods by microwave-assisted chemical-bath deposition for highly sensitive self-powered UV detection application. *Sens. Actuators A-Phys.* **2017**, *263*, 166–173. [CrossRef]
18. Han, C.; Chen, Z.; Zhang, N.; Colmenares, J.C.; Xu, Y.J. Hierarchically CdS decorated 1D ZnO nanorods-2D graphene hybrids: Low temperature synthesis and enhanced photocatalytic performance. *Adv. Funct. Mater.* **2015**, *25*, 221–229. [CrossRef]
19. Zhang, N.; Xie, S.J.; Wenig, B.; Xu, Y.J. Vertically aligned ZnO-Au@CdS core-shell nanorod arrays as an all-solid-state vectorial Z-scheme system for photocatalytic application. *J. Mater. Chem. A* **2016**, *4*, 18804–18814. [CrossRef]
20. Liu, S.Q.; Tang, Z.R.; Sun, Y.G.; Colmenares, J.C.; Xu, Y.J. One-dimension-based spatially ordered architectures for solar energy conversion. *Chem. Soc. Rev.* **2015**, *44*, 5053–5075. [CrossRef] [PubMed]
21. Wu, C.C.; Wuu, D.S.; Lin, P.R.; Chen, T.N.; Horng, R.H. Realization and manipulation of ZnO nanorod arrays on sapphire substrates using a catalyst-free metalorganic chemical vapor deposition technique. *J. Nanosci. Nanotechnol.* **2010**, *10*, 3001–3011. [CrossRef] [PubMed]
22. Wu, C.C.; Wuu, D.S.; Lin, P.R.; Chen, T.N.; Horng, R.H. Three-step growth of well-aligned ZnO nanotube arrays by self-catalyzed metalorganic chemical vapor deposition method. *Cryst. Growth Des.* **2009**, *9*, 4555–4561. [CrossRef]
23. Kawakami, M.; Hartanto, A.B.; Nakata, Y.; Okada, T. Synthesis of ZnO nanorods by nanoparticle assisted pulsed-laser deposition. *Jpn. J. Appl. Phys.* **2003**, *42*, L33–L35. [CrossRef]
24. Kim, M.S.; Nam, G.; Leem, J.Y. Photoluminescence studies of ZnO nanorods grown by plasma-assisted molecular beam epitaxy. *J. Nanosci. Nanotechnol.* **2013**, *13*, 3582–3585. [CrossRef] [PubMed]
25. Chen, Z.T.; Gao, L. A facile route to ZnO nanorod arrays using wet chemical method. *J. Cryst. Growth* **2006**, *293*, 522–527. [CrossRef]
26. Sardana, S.K.; Chandrasekhar, P.S.; Kumar, R.; Komarala, V.K. Efficiency enhancement of silicon solar cells with vertically aligned ZnO nanorod arrays as an antireflective layer. *Jpn. J. Appl. Phys.* **2017**, *56*, 040305. [CrossRef]
27. Ahsanylhaq, Q.; Umar, A.; Hahn, Y.B. Growth of aligned ZnO nanorods and nanopencils on ZnO/Si in aqueous solution: Growth mechanism and structural and optical properties. *Nanotechnology* **2007**, *18*, 115603. [CrossRef]
28. Xu, S.; Lao, C.S.; Weintraub, B.; Wang, Z.L. Density-controlled growth of aligned ZnO nanowire arrays by seedless chemical approach on smooth surfaces. *J. Mater. Res.* **2008**, *23*, 2072–2077. [CrossRef]

29. Wang, Z.; Qian, X.F.; Yin, J.; Zhu, Z.K. Aqueous solution fabrication of large-scale arrayed obelisk-like zinc oxide nanorods with high efficiency. *J. Solid State Chem.* **2004**, *177*, 2144–2149. [CrossRef]

30. Prabhu, M.; Mayandi, J.; Mariammal, R.N.; Vishnukanthan, V.; Pearce, J.M.; Soundararajan, N.; Ramachandran, K. Peanut shaped ZnO microstructures: Controlled synthesis and nucleation growth toward low-cost dye sensitized solar cells. *Mater. Res. Express* **2015**, *2*, 066202. [CrossRef]

31. Jang, J.M.; Kim, S.D.; Kim, S.D.; Choi, H.M.; Kim, J.Y.; Jung, W.G. Morphology change of self-assembled ZnO 3D nanostructures with different pH in the simple hydrothermal process. *Mater. Chem. Phys.* **2009**, *113*, 389–394. [CrossRef]

32. Wilson, S.J.; Hutley, M.C. The optical properties of 'moth eye' antireflection surfaces. *Opt. Acta* **1982**, *29*, 993–1009. [CrossRef]

33. Zhang, D.B.; Wang, S.J.; Cheng, K.; Dai, S.X.; Hu, B.B.; Han, X.; Shi, Q.; Du, Z.L. Controllable fabrication of patterned ZnO nanorod arrays: Investigations into the impacts on their morphology. *ACS Appl. Mater. Interfaces* **2012**, *4*, 2969–2977. [CrossRef] [PubMed]

34. Yue, S.S.; Lu, J.J.; Zhang, J.Y. Controlled growth of well-aligned hierarchical ZnO arrays by a wet chemical method. *Mater. Lett.* **2009**, *63*, 2149–2152. [CrossRef]

35. Bai, S.L.; Guo, T.; Zhao, Y.B.; Luo, R.X.; Li, D.Q.; Chen, A.F.; Liu, C.C. Mechanism enhancing gas sensing and first-principle calculations of Al-doped ZnO nanostructures. *J. Mater. Chem. A* **2013**, *1*, 11335–11342. [CrossRef]

36. Wang, M.J.; Shen, Z.R.; Chen, Y.L.; Zhang, Y.; Ji, H.M. Atomic structure-dominated enhancement of acetone sensing for a ZnO nanoplate with highly exposed (0001) facet. *CrystEngComm* **2017**, *19*, 6711–6718. [CrossRef]

37. Xu, J.P.; Shi, S.B.; Wang, C.; Zhang, Y.Z.; Liu, Z.M.; Zhang, X.G.; Li, L. Effect of surface-to-volume ratio on the optical and magnetic properties of ZnO nanorods by hydrothermal method. *J. Alloys Compd.* **2015**, *648*, 521–526. [CrossRef]

MDPI

St. Alban-Anlage 66

4052 Basel

Switzerland

Tel. +41 61 683 77 34

Fax +41 61 302 89 18

www.mdpi.com

Crystals Editorial Office

E-mail: crystals@mdpi.com

www.mdpi.com/journal/crystals